Electrical and
Electronic Principles
for Technicians
Volume 2

Electrical and

Longman Scientific & Technical

Longman Scientific & Technical,
Longman Group UK Limited,
Longman House, Burnt Hill, Harlow,
Essex, CM20 2JE, England
and Associated Companies throughout the world.

© Longman Group UK Limited 1992

First published 1992

British Library Cataloguing in Publication Data
A catalogue record for this title is available from the
British Library

ISBN 0-582-08055-X

Set in Compugraphic Times 10/12 pt.

Printed in Malaysia by PMS

Contents

Preface vii

1 **Single-phase circuits 1**
The series circuit 1
The parallel circuit 7
The series—parallel circuit 14
Resonant circuit 15
Power factor and power factor correction 22

2 **Three-phase circuits 27**
The star connection 31
The delta connection 36
Relative merits of the star and delta connections 38
Unbalanced loads 38
Measurement of three-phase power 41

3 **Circuit theorems 46**
Current and voltage sources 46
Thevenin's theorem 47
Norton's theorem 50
The superposition theorem 54
Maximum power transfer theorem 57

4 **D.C. transients 61**
Resistance—capacitance circuits 62
Inductance—resistance circuits 72
The effect of circuit time constant on rectangular
    waveforms 75
Integrating and differentiating circuits 78
Appendix A Voltage growth and current decay in an $R-C$
    circuit 81
Appendix B Voltage and current decay in an $R-C$
    circuit 81
Appendix C Time constant 82
Appendix D Current growth in an inductive circuit 82
Appendix E Current decay in an inductive circuit 83

5 **Electrical Machines 84**
The a.c. generator 87

The d.c. generator 89
The d.c. motor 97
The induction motor 105
Small motors 114
The stepper motor 115

6 **Decibels, attenuation and filters 118**
The decibel 119
Attenuation and attenuators 123
Filters 126

7 **Modulation 133**
Amplitude modulation 134
Frequency modulation 142
The relative merits of amplitude and frequency
    modulation 145
Phase modulation 147
Pulse modulation 148
Appendix A 155

8 **Control systems 156**
Open-loop and closed-loop systems 156
Closed-loop control systems 158
Computer or microprocessor control 173
Relative merits of analogue and digital control 178
Instability in a control system 179

9 **Measurements 183**
The loading effect of a voltmeter 185
Frequency effects on voltmeters 186
Waveform errors with voltmeters 187
The decibelmeter 189
A.C. bridges 191
The transformer ratio-arm bridge 196
The Q meter 199
The cathode ray oscilloscope 200
Instruments for digital measurements 204
Appendix A 211

Exercises 212

Answers to numerical exercises 218

Index 223

Preface

The Business and Technician Education Council (BTEC) programmes in electrical/electronic/communication/computer engineering require some study of electrical and electronic principles at both level II and level III. The BTEC bank for electrical and electronic principles is broadly divided into level II and level III topics. The preceding volume, *Electrical and Electronic Principles for Technicians* Vol. 1, covered the level II part of the BTEC bank and this volume covers the level III material. Together the two books provide complete coverage of the BTEC requirements in electrical and electronic principles for technicians.

The book has been written on the assumption that the reader will have already studied both electrical principles and mathematics at the level II standard. In one or two places a knowledge of the operational amplifier, or op-amp, is also assumed.

A large number of worked examples are provided throughout the book and exercises on each chapter are to be found at the back of the book. A worked solution to each numerical exercise is given.

D.C.G.

1 Single-phase circuits

A single-phase a.c. circuit consists of a number of components, such as resistors, capacitors and inductors, connected either in series or in parallel, or perhaps a series−parallel combination. The calculation of the current(s) flowing, and the power(s) dissipated in an a.c. circuit when an a.c. voltage is applied to it, can be carried out in various ways. If the applied voltage is of sinusoidal waveform and all the components are linear, i.e. obey Ohm's law, the solution of the circuit can be carried out with the aid of a phasor diagram.

The reactance X_L of an inductor, or the reactance X_C of a capacitor, represents the opposition of the component to the flow of a sinusoidal a.c. current. Inductive reactance X_L is equal to $2\pi fL$ ohms, and capacitive reactance is equal to $1/2\pi fC$ ohms, where f is the frequency of the sinusoidal applied voltage.

When the phasor diagram of a circuit is drawn the reference phasor is normally chosen to be the phasor that represents the common quantity. This means that for a series circuit the current phasor is selected to be the reference phasor, and for a parallel circuit the voltage phasor is chosen as the reference.

The series circuit

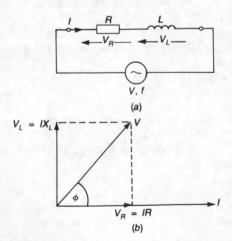

Fig. 1.1 (a) Series RL circuit; (b) phasor diagram of (a)

Resistance and inductance

Figure 1.1(a) shows a series a.c. circuit that consists of a resistor R and an inductor L connected in series with one another and with a voltage source of e.m.f. V volts and frequency f hertz. The inevitable self-resistance of the inductor is assumed to be included in with the resistance R. The current I that flows in the circuit is determined by the applied voltage and the *impedance* Z of the circuit, i.e. $I = V/Z$. The voltage V_R dropped across the resistor R is equal to the product of the current and the resistance, IR, and it is in phase with the applied voltage V. The voltage V_L developed across the inductance is equal to the product of the current and the inductive reactance, IX_L, and it leads the applied voltage by 90°. The phasor diagram for the circuit is shown by Fig. 1.1(b). The magnitude of the source voltage is, from Fig. 1.1(b), given by

$$
\begin{aligned}
|V| &= \sqrt{(V_R^2 + V_L^2)} \\
&= \sqrt{[(IR)^2 + (I\omega L)^2]} \\
&= I\sqrt{(R^2 + \omega^2 L^2)}
\end{aligned}
$$

The magnitude of the impedance of the circuit is

$$|Z| = |V|/I = \sqrt{(R^2 + \omega^2 L^2)} \; \Omega. \tag{1.1}$$

The supply voltage leads the current by angle ϕ, where

$$\phi = \tan^{-1}(\omega L/R) \tag{1.2}$$

Since impedance is equal to V/I the circuit impedance has the same phase angle and hence

$$Z = \sqrt{(R^2 + \omega^2 L^2)} \; \angle \tan^{-1}(\omega L/R) \; \Omega \tag{1.3}$$

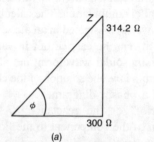

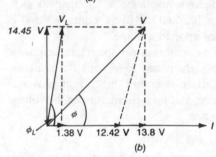

(a)

(b)

Fig. 1.2

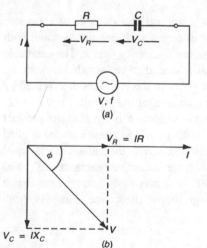

(a)

(b)

Fig. 1.3   (a) Series RC circuit.; (b) phasor diagram of (a)

Example 1.1

A 10 mH inductor has a self-resistance of 30 Ω and is connected in series with a 270 Ω resistor. The circuit is connected across a 20 V, 5 kHz supply. (a) Draw the impedance triangle of the circuit and use it to calculate the impedance of the circuit. (b) Calculate the current that flows in the circuit. (c) Draw the phasor diagram of the circuit and then calculate the voltage dropped across the inductor. (d) Calculate the power dissipated in the circuit.

Solution
(a) The reactance of the inductor is $X_L = 2\pi \times 5000 \times 10 \times 10^{-3} = 314.2 \; \Omega$.
The impedance triangle is shown in Fig. 1.2(a). From the triangle
$$|Z| = \sqrt{(300^2 + 314.2^2)} = 434.4\Omega$$
and phase angle $\phi = \tan^{-1}(314.2/300) = 46.3°$. Therefore,
$$Z = 434.4 \angle 46.3° \; \Omega. \quad (Ans.)$$
(b) Current in circuit = $V/Z = 20/434.4 \angle 46.3° = 46 \angle -46.3°$ mA.
$$(Ans.)$$
But taking the current to provide the reference phasor $I = 46$ mA.   (*Ans.*)
(c) The voltage dropped across the 10 mH inductor is
$$V_1 = IX_L = 46 \times 10^{-3} \times 314.2 = 14.45 \angle 90° \; V.$$
The voltage dropped across the 30 Ω self-resistance of the inductor is
$$I_r = 46 \times 10^{-3} \times 30 = 1.38 \, V.$$
The voltage across the 270 Ω resistor is $46 \times 10^{-3} \times 270 = 12.42$ V.
The phasor diagram is shown in Fig. 1.2(b). From the diagram, the voltage V_L across the inductor is
$$V_L = \sqrt{(1.38^2 + 14.45^2)} \; \angle \tan^{-1}(14.45/1.38)$$
$$= 14.52 \angle 84.5° \; V. \quad (Ans.)$$
(*Note*: $\sqrt{(14.45^2 + 13.8^2)} = 20$ V.)
(d) Power dissipated = $|I|^2 R = (46 \times 10^{-3})^2 \times 300 = 635$ mW.
$$(Ans.)$$
$$= VI \cos \phi = 20 \times 46 \times 10^{-3} \times \cos 46.3°$$
$$= 636 \; mW. \quad (Ans.)$$

Resistance and capacitance

Figure 1.3(a) shows a capacitor C and a resistor R connected in series across a voltage source of e.m.f. V volts and frequency f hertz. Since the voltage across a capacitor lags the current by 90° the phasor diagram of the circuit is as shown by Fig. 1.3(*b*).

The magnitude of the applied voltage V is given by the phasor sum of V_R and V_C, i.e.

$$|V| = \sqrt{(V_R^2 + V_C^2)} = \sqrt{[(IR)^2 + (IX_C)^2]}$$
$$= I\sqrt{[R^2 + (1/\omega C)^2]}$$

The phase angle ϕ between the applied voltage and the current flowing in the circuit is

$$\phi = \tan^{-1}(V_C/V_R) = \tan^{-1}(X_C/R) = \tan^{-1}(1/\omega CR)$$

The impedance Z of the circuit is equal to the ratio voltage/current, and hence

$$Z = V/I = \sqrt{(R^2 + 1/\omega^2 C^2)} \angle \tan^{-1}(1/\omega CR).\ \Omega \qquad (1.4)$$

Example 1.2

A series a.c. circuit consists of a 270 Ω resistor and a 2.2 μF capacitor connected across a 12 V, 796 Hz voltage source. (*a*) Draw the impedance triangle of the circuit. (*b*) Calculate the current flowing and the power dissipated in the circuit. (*c*) Calculate the frequency at which the current will lead the applied voltage by (i) 45° and (ii) 20°.

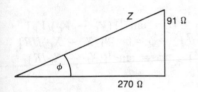

Fig. 1.4

Solution
(*a*) The reactance of the capacitor is
$$X_C = 1/(2\pi \times 796 \times 2.2 \times 10^{-6}) = 90.9\ \Omega \simeq 91\ \Omega.$$
Figure 1.4 shows the impedance triangle and from this
$$Z = \sqrt{(270^2 + 91^2)} \angle \tan^{-1}(91/270) = 284.9 \angle 18.6°\ \Omega.$$
(*b*) $|I| = V/|Z| = 12/284.9 = 42.12$ mA.   (*Ans.*)
Power $P = |I|^2 = (42.12 \times 10^{-3})^2 \times 270 = 479$ mW.   (*Ans.*)
or $P = V|I|\cos 18.6° = 479$ mW.   (*Ans.*)
(*c*) (i) $\tan 45° = 1 = X_C/R = 1/\omega CR$. Therefore
$$f = 1/(2\pi \times 270 \times 2.2 \times 10^{-6}) = 268\ \text{Hz}. \quad (Ans.)$$
(ii) $\tan 20° = 0.364 = 1/\omega CR$, and
$$f = 1/(2\pi \times 270 \times 2.2 \times 10^{-6} \times 0.364) = 736\ \text{Hz}. \quad (Ans.)$$

Capacitance, inductance and resistance

When a capacitor, an inductor and a resistor are connected in series across a voltage source of e.m.f. V volts and frequency f hertz (see Fig. 1.5), the current I which flows in the circuit is determined by the impedance, Z, of the circuit. At any frequency f the effective reactance X_T of the circuit is equal to the difference between the reactances, X_C and X_L, of the capacitor and the inductor respectively. The phasor diagram of the circuit is shown in Fig. 1.6. The current phasor is taken as the reference and the voltage V_R across the resistor is in phase with it. The phasor that represents the capacitor voltage

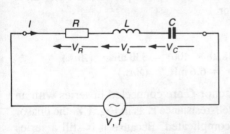

Fig. 1.5 Series *RLC* circuit

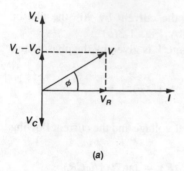

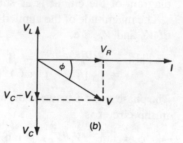

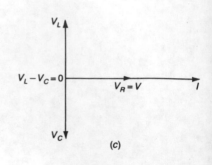

(a)　　　(b)　　　(c)

Fig. 1.6 Phasor diagram of a series RLC circuit: (a) $X_L > X_C$; (b) $X_C > X_L$; and (c) $X_L = X_C$

V_C lags the current phasor by 90°, and the phasor representing the voltage V_L leads the current phasor by 90°.

Three possible cases exist: (a) the reactance of the inductor is greater than the reactance of the capacitor, then V_L is greater than V_C as shown by Fig. 1.6(a); (b) $V_C > V_L$, shown in Fig. 1.6(b); and (c) $V_C = V_L$ shown by Fig. 1.6(c). When $V_L > V_C$ the applied voltage V is equal to the phasor sum of V_R and the difference between V_L and V_C, i.e. $V_L - V_C$. Thus

$$|V| = \sqrt{[V_R^2 + (V_L - V_C)^2]} \quad \phi = \tan^{-1}[(V_L - V_C)/V_R]$$
$$= \sqrt{[(IR)^2 + I^2(X_L - X_C)^2]} \quad \phi = \tan^{-1}[(X_L - X_C)/IR]$$
$$= I\sqrt{[R^2 + (X_L - X_C)^2]} \quad \phi = \tan^{-1}[(X_L - X_C/R]$$

The impedance Z of the circuit is

$$Z = V/I = \sqrt{[R^2 + (X_L - X_C)^2]} \angle \tan^{-1}(X_L - X_C)/R \ \Omega \quad (1.5)$$
$$Z = \sqrt{[R^2 + (\omega L - 1/\omega C)^2]} \angle \tan^{-1}(\omega L - 1/\omega C)/R \ \Omega \quad (1.6)$$

Example 1.3

If, in the circuit shown in Fig. 1.7, $V_C = V_R = V_L/2$ calculate the values of the capacitor and the inductor.

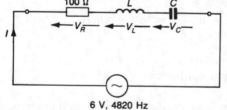

Fig. 1.7

Solution
$100I = I/\omega C$, $C = 1/(2\pi \times 4820 \times 100) = 330$ nF.　(*Ans.*)
$100I = \omega LI/2$, $L = 100/4820\pi = 6.6$ mH.　(*Ans.*)

When a resistor R and a capacitor C are connected in series with an inductor of inductance L and self-resistance r, as in Fig. 1.8, the phasor diagram is somewhat more complicated. Because it is still a series circuit the current phasor is taken as the reference. The voltage Ir across the self-resistance of the inductor is in phase with the current and the voltage IX_L across the inductive reactance X_L leads the current by 90°. The phasor sum of these two voltages gives the voltage V_L across the inductor; this is shown by Fig. 1.9(a). The voltage V_C across the capacitor lags the current by 90° and this is shown by Fig. 1.9(b). It has been assumed that $IX_L > IX_C$ so that the phasor sum,

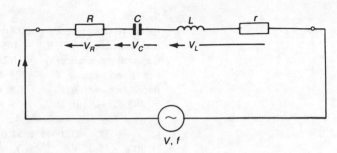

Fig. 1.8 A practical inductor in series with a resistor and a capacitor

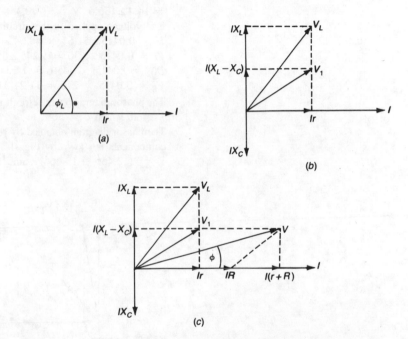

Fig. 1.9 Phasor diagram of the circuit shown in Fig. 1.8: (a) for V_L (b) for V_L plus V_C; (c) for the complete circuit

V_1, of V_L and V_C leads the current. If, of course, $IX_C > IX_L$ the phasor for V_1 would lag the current phasor. When $IX_C = IX_L$ the current will be in phase with the applied voltage and this is known as the resonant condition and it is dealt with on p. 15. Lastly, the voltage $V_R = IR$ dropped across the resistor R is in phase with the current and the phasor sum of V_R and V_1 is equal to the supply voltage V. This is shown by Fig. 1.9(c). The supply voltage is also equal to the phasor sum of the voltages $I(X_L - X_C)$ and $I(r + R)$.

Example 1.4

A 55 mH coil has a self-resistance of 30 Ω and it is connected in series with a 220 nF capacitor and a 270 Ω resistor. The circuit is connected to a 20 V voltage source at a frequency of 1000 Hz. Calculate (a) the voltage across the capacitor and (b) across the coil, and draw the phasor diagram for the circuit.

Solution

Inductive reactance $X_L = 2000\pi \times 55 \times 10^{-3} = 345.6\,\Omega$.

Coil impedance $= \sqrt{(30^2 + 345.6^2)} \angle \tan^{-1}(345.6/30) = 346.9 \angle 85°\,\Omega$.

Capacitive reactance $X_C = 1/(2\pi \times 1000 \times 220 \times 10^{-9}) = 723.4\,\Omega$.

Effective reactance $X_T = (723.4 - 345.6) = 377.8\,\Omega$ capacitive.

Impedance $= \sqrt{(300^2 + 377.8^2)} \angle \tan^{-1}(-377.8/300)$

$= 482.4 \angle -51.6°\,\Omega$.

Current $I = V/Z = 20/482.4 \angle -51.6° = 41.5 \angle 51.6°\,mA$.

(a) $V_C = IX_C = 0.0415 \angle 51.6° \times 723.4 \angle -90°$

$= 30 \angle -38.4°\,V$.   (*Ans.*)

(b) $V_L = IZ_L = 0.0415 \angle 51.6° \times 347 \angle 85°$

$= 14.4 \angle 136.6°\,V$.   (*Ans.*)

The voltage applied to the circuit is

$IZ = 0.0415 \angle 51.6° \times 482.4 \angle -51.6° = 20 \angle 0°\,V$.

$Vr = 0.0415 \times 30 = 1.245 \angle 0°\,V$.

$IX_L = 0.0415 \times 345.6 = 14.34 \angle 90°\,V$.

$V_R = 0.0415 \times 270 = 11.21 \angle 0°\,V$.

The phasor diagram of the circuit is shown in Fig. 1.10(*a*). By measurement $V = 20\,V$ and $\phi = 50°$.

The phasor diagram obtained by resolving V_L into its horizontal and vertical components is shown by Fig. 1.10(*b*). From this

$V = \sqrt{(12.45^2 + 15.66^2)} \angle \tan^{-1}(-15.66/12.45) = 20 \angle -51.5°\,V$.

(*Ans.*)

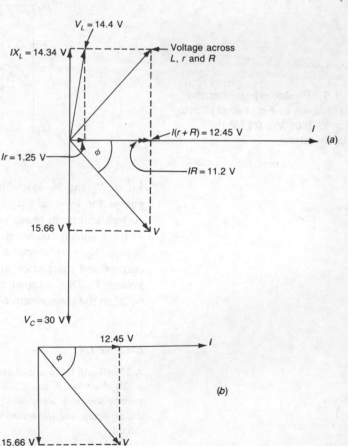

Fig. 1.10

The parallel circuit

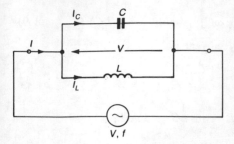

Fig. 1.11 A pure inductor in parallel with a capacitor

Pure inductance and capacitance

Figure 1.11 shows a pure inductor, i.e. one having zero self-resistance, connected in parallel with a capacitor. A voltage source of e.m.f. V volts and frequency f hertz is connected across the circuit. The two components have the same voltage V across them and so the currents in the two branches of the circuit are

$$I_C = V/X_C = V\omega C \quad \text{and} \quad I_L = V/X_L = V/\omega L.$$

The capacitor current I_C leads the applied voltage by 90° and inductor current I_L lags the applied voltage by 90°. The two currents are in anti-phase with one another and hence the total current I is equal to the difference between I_L and I_C. The three possible phasor diagrams for the circuit are shown in Figs 1.12(a)–(c). At low frequencies the reactance of the capacitor is higher than the inductive reactance and so $I_L > I_C$ (Fig. 1.12(a)); at high frequencies the opposite is true and so $I_C > I_L$ (Fig. 1.12(b)); lastly, at one particular frequency the reactances of the two components are equal to one another and so therefore are the currents that flow in them. The phasor diagram for this case, known as *resonance*, is given in Fig. 1.12(c). Resonance is considered further on p. 19.

Consider Fig. 1.12(a); the supply current is

$$I = I_L - I_C$$
$$= V[1/(\omega L) - \omega C] \tag{1.7}$$

The *admittance* Y of the circuit is the ratio I/V siemens (S) and hence

$$Y = I/V = (1/\omega L - \omega C) \text{ S} \tag{1.8}$$

The *susceptance* B_L of the inductor is equal to $1/\omega L$ and the *susceptance* B_C of the capacitor is equal to ωC. Clearly, susceptance is the reciprocal of reactance. Inductive reactance is positive and capacitive reactance is negative and so inductive susceptance is negative and capacitive susceptance is positive.

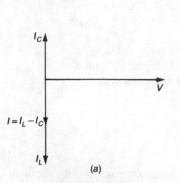

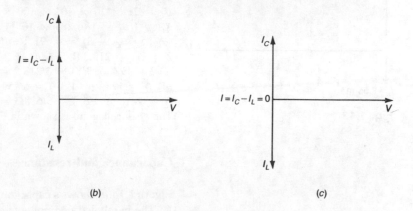

Fig. 1.12 Phasor diagram of parallel LC circuit: (a) $X_C > X_L$; (b) $X_L > X_C$; (c) $X_C = X_L$

The *impedance Z* of the circuit is the reciprocal of the admittance, i.e.

$$Z = 1/Y = 1/[1/(\omega L) - \omega C] = \omega L/(1 - \omega^2 LC) \; \Omega. \qquad (1.9)$$

In Fig. 1.12(*b*) the supply current is

$$I = I_C - I_L = V[\omega C - 1/(\omega L)]$$

the admittance is

$$Y = I/V = \omega C - 1/\omega L = (\omega^2 LC - 1)/\omega L.$$

Hence the impedance is

$$Z = 1/Y = \omega L/(\omega^2 LC - 1)$$

Lastly, in Fig. 1.12(*c*), $I = I_L - I_C = 0$ and this means that the frequency is such that $\omega_0 C = 1/(\omega_0 L)$.

In this ideal case, when the inductor has zero self-resistance,

$$\omega_0^2 = 1/LC,$$

or

$$f_0 = 1/2\pi \sqrt{LC} \; \text{Hz}. \qquad (1.10)$$

as for a series-resonant circuit.

Example 1.5

A 10 mH inductor of negligible self-resistance is connected in parallel with a 100 nF capacitor and the parallel circuit is connected to a 6 V voltage source at 10 kHz. Calculate (*a*) the reactance and the susceptance of each component, (*b*) the admittance and the impedance of the circuit, (*c*) the current flowing into the circuit, and (*d*) the current flowing in each component. Draw the phasor diagram for the circuit.

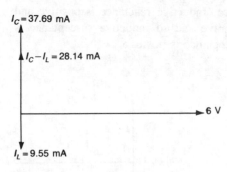

$I_C = 37.69$ mA

$I_C - I_L = 28.14$ mA

6 V

$I_L = 9.55$ mA

Fig. 1.13

Solution
(*a*) $X_L = 2\pi \times 10 \times 10^3 \times 10 \times 10^{-3} = 628.3 \; \Omega.$ (*Ans.*)
$B_L = 1/X_L = 1/628.3 = 1.592 \times 10^{-3} \, \text{S}.$ (*Ans.*)
$B_C = 2\pi \times 10 \times 10^3 \times 100 \times 10^{-9} = 6.283 \times 10^{-3} \, \text{S}.$ (*Ans.*)
$X_C = 1/B_C = 1/(6.283 \times 10^{-3}) = 159.2 \; \Omega.$ (*Ans.*)
(*b*) $Y = B_C - B_L = 4.691 \times 10^{-3} \, \text{S}.$ (*Ans.*)
$Z = 1/Y = 213.2 \; \Omega.$ (*Ans.*)
(*c*) $I = V/Z \text{ (or } VY) = 6/213.2 = 28.14 \, \text{mA}.$ (*Ans.*)
(*d*) $I_C = V/X_C \text{ (or } VB_C) = 6/159.2 = 37.69 \, \text{mA}.$ (*Ans.*)
$I_L = V/X_L \text{ (or } VB_L) = 6/628.3 = 9.55 \, \text{mA}.$ (*Ans.*)
The phasor diagram is shown in Fig. 1.13.

Capacitance and resistance

Figure 1.14(*a*) shows a capacitor *C* connected in parallel with a resistor *R*. The paralleled components are connected to a voltage source of e.m.f. *V* volts and frequency *f* hertz. The current I_R flowing in the resistor is equal to *V/R* and it is in phase with the applied voltage.

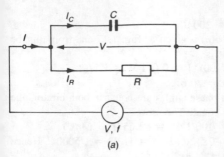

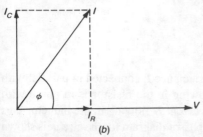

Fig. 1.14 (a) A capacitor in parallel with a resistor; (b) phasor diagram of (a)

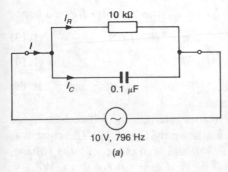

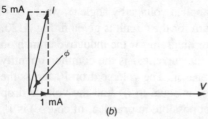

Fig. 1.15

Fig. 1.16

The current I_C in the capacitor is equal to V/X_C and it leads the applied voltage by 90°. The phasor diagram for the circuit is shown by Fig. 1.14(b). From the phasor diagram the magnitude of the supply current is

$$|I| = \sqrt{(I_R^2 + I_C^2)} = \sqrt{[(V/R)^2 + (V\omega C)^2]}$$
$$= V\sqrt{[(1/R)^2 + (V\omega C)^2]}$$

The phase angle ϕ between the applied voltage V and the supply current I is

$$\phi = \tan^{-1}(I_C/I_R) = \tan^{-1}(\omega CR)$$

The admittance Y of the circuit is

$$Y = I/V = \sqrt{[(1/R)^2 + \omega^2 C^2]} \, \angle \tan^{-1}\omega CR \text{ S.} \qquad (1.11)$$

or

$$Y = \sqrt{(G^2 + B_C^2)} \, \angle \tan^{-1}(B_C/G) \text{ S.} \qquad (1.12)$$

where G is the conductance of the resistor and B_C is the susceptance of the capacitor.

The impedance Z of the circuit is the reciprocal of the circuit's impedance, i.e. $Z = 1/Y$.

Example 1.6

For the circuit given in Fig. 1.15(a) calculate (a) the current taken from the voltage source and (b) the admittance and the impedance of the circuit.

Solution
(a) $B_C = 2\pi \times 796 \times 0.1 \times 10^{-6} = 5 \times 10^{-4}$ S.
$I_C = 10 \times 5 \times 10^{-4} = 5$ mA.
$I_R = 10/(10 \times 10^3) = 1$ mA.
The phasor diagram is shown in Fig. 1.15(b) and from this the supply current is

$$I = \sqrt{(1^2 + 5^2)} \, \angle \tan^{-1}(5/1) = 5.1 \angle 78.7° \text{ mA.} \quad \text{(Ans.)}$$

(b) The admittance $Y = I/V = 0.51 \angle 78.7°$ mS. (Ans.)
$Z = 1/Y = 1960.8 \angle -78.7°$ Ω. (Ans.)

Example 1.7

The values of the supply current and the applied voltage and the phase difference between them are the same for each of the circuits shown in Figs 1.16(a) and (b). Calculate the values of R and C if the frequency of the supply is $5000/2\pi$ Hz.

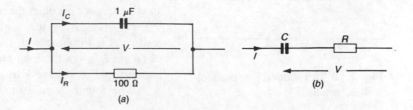

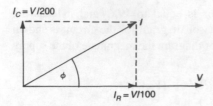

Fig. 1.17

Solution

$X_C = 1/(5000 \times 1 \times 10^{-6}) = 200\,\Omega$.

The phasor diagram for Fig. 1.16(a) is given by Fig. 1.17 and from this the magnitude of the supply current is

$$|I| = V \sqrt{[(1/200)^2 + (1/100)^2]} = 0.011\,V = 1.1I_R.$$

The angle $\phi = \tan^{-1}(100/200) = 26.6°$.

Since the currents, voltages and phase angles are equal in both circuits the powers dissipated must also be equal. Therefore

$$100I_R^2 = (1.1I_R)^2 R \text{ or } R = 100/1.1^2 = 82.65\,\Omega. \quad (Ans.)$$

Tan $26.6° = 0.5 = X_C/R$, $X_C = 0.5 \times 82.65 = 41.33 = 1/5000C$. Hence

$$C = 1/(41.33 \times 5000) = 4.84\,\mu F. \quad (Ans.)$$

Inductance and resistance

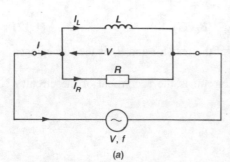

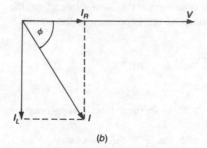

Fig. 1.18 (a) A pure inductor in parallel with a resistor; (b) phasor diagram of (a)

Figure 1.18(a) shows a pure inductance L connected in parallel with a resistor R. The current I_R flowing in the resistor is in phase with the applied voltage V, and the current I_L flowing in the inductor lags the applied voltage by 90°. The phasor diagram for the circuit is shown by Fig. 1.18(b). The current I supplied to the circuit is the phasor sum of I_R and I_L, i.e.

$$I = \sqrt{(I_R^2 + I_L^2)} \; \angle \tan^{-1}(I_L/I_R)$$
$$= V \sqrt{[(1/R)^2 + (1/\omega L)^2]} \; \angle \tan^{-1}(R/\omega L)$$

The admittance Y of the circuit is equal to the ratio I/V and hence

$$Y = \sqrt{[(1/R)^2 + (1/\omega L)^2]} \; \angle \tan^{-1}(R/\omega L) \text{ S.} \quad (1.13)$$
$$= \sqrt{(G^2 + B_L^2)} \; \angle \tan^{-1}(B_L/R) \text{ S.} \quad (1.14)$$

where G is the conductance $(1/R)$ of the resistor and B_L is the susceptance $(1/X_L)$ of the inductor.

In practice, the inductor will always have some self-resistance and so the circuit will really be as shown by Fig. 1.19. The current I_R in the resistor is still equal to V/R and is in phase with the voltage, but now the current I_L in the inductor is equal to $V/Z_L = V/\sqrt{(r^2 + \omega^2 L^2)}$. I_L will lag the applied voltage by angle ϕ, where $\phi = \tan^{-1}(\omega L/r)$. The phasor diagram for the circuit is given in Fig. 1.20. Figure 1.20(a) gives the phasor diagram for the inductive branch on its own; since for this branch the current I_L is the common quantity it is taken to be the reference phasor. The voltage drop V_r across the self-resistance r of the inductor is equal to $I_L r$ and the voltage drop across the pure inductance (not possible in practice, of course) is $V_L = I_L X_L$. The applied voltage is the phasor sum of V_r and V_L. For the complete circuit the applied voltage V becomes the common quantity and so its phasor becomes the reference. This is shown by Fig. 1.20(b) where I_L lags V by the angle ϕ_L. The current I_R flowing in the resistor R is in phase with the applied voltage and the phasor sum of I_R and I_L gives the total current I into the circuit. The current I lags the applied voltage V by angle ϕ as shown. The values of I and ϕ can be found by measurement if the phasor diagram is drawn

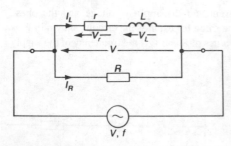

Fig. 1.19 An inductor in parallel with a resistor

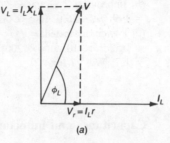

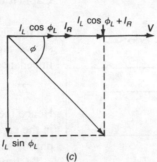

Fig. 1.20 Phasor diagram for the circuit shown in Fig. 1.19

accurately enough but it is more accurate to obtain their values by calculation. To do this the current I_L should be resolved into its horizontal and vertical components, $I_L \cos \phi_L$ and $I_L \sin \phi_L$ respectively, and this has been done in Fig. 1.20(c). From this diagram

$$|I| = \sqrt{[(I_R + I_L \cos \phi_L)^2 + (I_L \sin \phi_L)^2]}$$

and

$$\phi = \tan^{-1}[(I_L \sin \phi_L)/(I_R + I_L \cos \phi_L)]$$

Example 1.8

A 1 H inductor has a resistance of 200 Ω and it is connected in parallel with a 330 Ω resistor. Draw the phasor diagram for the circuit and calculate (a) the current that flows into the circuit, (b) the impedance of the circuit, and (c) the power dissipated, when 20 V at 50 Hz is applied.

Solution

$X_L = 100\pi = 314.2 \,\Omega$.

$Z_L = \sqrt{(200^2 + 314.2^2)} \angle \tan^{-1}(314.2/200) = 372.5 \angle 57.5° \,\Omega$.

$I_L = 20/372.5 \angle 57.5° = 53.7 \angle -57.5° \,\text{mA}$.

$V_L = I_L X_L = (53.7 \times 10^{-3} \times 314.3) \angle(-57.5 + 90)° = 16.9 \angle 32.5° \,\text{V}$.

$V_r = I_L r = (53.7 \times 10^{-3} \times 200) \angle -57.5° = 10.74 \angle -57.5° \,\text{V}$.

$I_R = 20/330 = 60.6 \,\text{mA}$.

(a) The phasor diagram is shown in Fig. 1.21(a). Resolving I_L into its horizontal and vertical components gives $53.7 \cos 57.5° = 28.9 \,\text{mA}$, and $53.7 \sin 57.5° = 45.3 \,\text{mA}$ respectively. The resulting phasor diagram is shown in Fig. 1.21(b), from which the magnitude of the current is

$$|I| = \sqrt{(89.5^2 + 45.3^2)} = 100.3 \,\text{mA}$$

and the phase angle $\phi = \tan^{-1}(45.3/89.5) = 26.9°$. Therefore

$$I = 100.3 \angle 26.9° \,\text{mA}. \quad (Ans.)$$

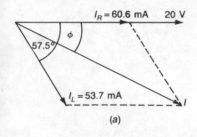

(a)

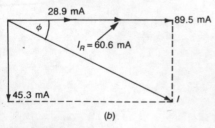

(b)

Fig. 1.21

(b) Impedance $Z = V/I = 20/0.1004 \angle 26.9°$
$= 199.4 \angle -26.9° \, \Omega.$ (*Ans.*)
(c) Power dissipated $= I_L^2 r + I_R^2 R$
$= [(53.7 \times 10^{-3})^2 \times 200] + [(60.6 \times 10^{-3})^2 \times 330]$
$= 1.79 \, \text{W}.$ (*Ans.*)

Capacitance and inductance

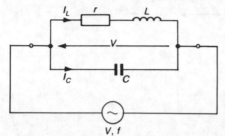

Fig. 1.22 An inductor in parallel with a capacitor

All practical inductors possess inevitable self-resistance and Fig. 1.22 shows an inductor connected in parallel with a capacitor. The impedance Z_L of the inductive branch of the circuit is

$$Z_L = \sqrt{(r^2 + \omega^2 L^2)} \ \angle \tan^{-1}(\omega L/R)$$

Hence the current I_L in this branch is $I_L = V/Z_L$ and the current I_C in the capacitor is $V/X_C = V\omega C$. The supply current is equal to the phasor sum of I_C and I_L.

To draw the phasor diagram of the circuit start with the inductive branch; the phasor diagram for this is shown in Fig. 1.23(*a*). The current I_C leads the voltage V by 90° and its phasor has been added to the phasor diagram to give Fig. 1.23(*b*). This diagram has been orientated to make the voltage phasor the reference phasor. The supply current I is the phasor sum of I_C and I_L and this is shown by the final phasor diagram given in Fig. 1.23(*c*). The phase angle ϕ between

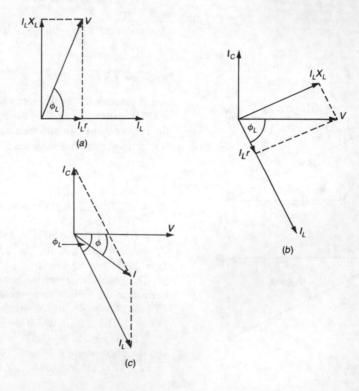

Fig. 1.23 Phasor diagram for the circuit shown in Fig. 1.22

the applied voltage and the supply current may be either a leading or a lagging (as shown) angle, depending upon the relative values of the currents I_C and I_L. This is, of course, a condition that changes with frequency. When the frequency is such that the phase angle ϕ is zero so that the applied voltage and the supply current are in phase with one another the circuit is said to be *resonant*.

Example 1.9

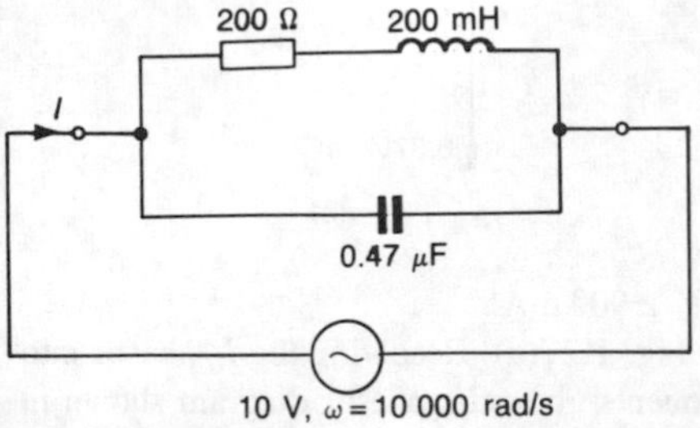

For the circuit given in Fig. 1.24 calculate (a) the current that flows in the inductor, (b) the supply current, and (c) the impedance of the circuit.

Solution

$X_L = 10 \times 10^{-3} \times 20 \times 10^{-3} = 200\,\Omega$.

$X_C = 1/(10 \times 10^3 \times 0.47 \times 10^{-6}) = 213\,\Omega$.

(a) $Z_L = \sqrt{(200^2 + 200^2)} \angle \tan^{-1}(200/200) = 282.8 \angle 45°\,\Omega$.

$I_L = 10/282.8 \angle 45° = 35.36 \angle -45°\,\text{mA}$.   (*Ans.*)

(b) $I_C = V/X_C = 10/213 = 46.95 \angle 90°\,\text{mA}$.

The phasor diagram for the circuit is shown by Fig. 1.25(a). Resolving the phasor I_L into its horizontal and vertical components gives

$\qquad 35.36 \cos -45° = 35.36 \sin -45° = 25\,\text{mA}$ (see Fig. 1.25(b)).

From Fig. 1.25(b) $|I| = \sqrt{(25^2 + 21.95^2)} = 33.27\,\text{mA}$

and the phase angle $\phi = \tan^{-1}(21.95/25) = 41.3°$. Hence the supply current is

$\qquad I = 33.27 \angle 41.3°\,\text{mA}$.   (*Ans.*)

(c) $Z = V/I = 10/(33.27 \times 10^{-3}) \angle 41.3° = 300.6 \angle -41.3\,\Omega$.   (*Ans.*)

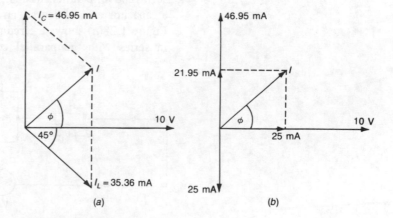

Fig. 1.25
(a)   (b)

Example 1.10

Determine the voltage across the circuit shown in Fig. 1.26.

Solution

$X_L = 2\pi \times 1000 \times 20 \times 10^{-3} = 125.7\,\Omega$.

$X_C = 1/(2\pi \times 1000 \times 1 \times 10^{-6}) = 159.2\,\Omega$.

$Z_L = \sqrt{(50^2 + 125.7^2)} \angle \tan^{-1}(125.7/50) = 135.3 \angle 68.3°\,\Omega$.

$I_L = V/135.3 \angle 68.3° = 7.39V \angle -68.3\,\text{mA}$.

Fig. 1.24

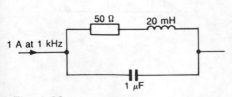

Fig. 1.26

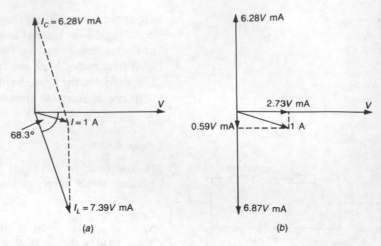

Fig. 1.27

(a)

(b)

$I_C = V/159.2 \angle -90° = 6.28V \angle 90°$ mA.

The phasor diagram is shown in Fig. 1.27(a). Resolving the I_L phasor into its horizontal and vertical components gives the phasor diagram shown in Fig. 1.27(b) since

$7.39V \cos 68.3° = 2.73V$ mA and $7.39V \sin 68.3° = 6.87V$ mA

From Fig. 1.27(b) $I = 1 = [\sqrt{(2.73^2 + 0.59^2)} \times 10^{-3}] V,$

$1 = 2.793 \times 10^{-3}V$, and $V = 358$ V. (*Ans.*)

The series–parallel circuit

The number of possible series–parallel circuits is very large and can only be solved using a suitable combination of the methods used for series and for parallel circuits. The practical inductor in parallel with a capacitor was, of course, an example of a series–parallel circuit. Figure 1.28(a) shows a circuit in which a resistor R_1 is connected in series with the parallel combination of a resistor R_2 and a

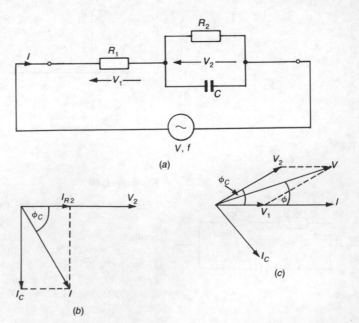

(a)

Fig. 1.28 (a) A series–parallel circuit; (b) and (c) phasor diagrams for (a)

(b)

(c)

capacitor C. The parallel branch of the circuit should first be considered and the voltage V_2 across it taken as the reference phasor. The phasor diagram for this part of the circuit is shown by Fig. 1.28(b).

The phasor sum of the currents I_{R_2} and I_C gives the supply current I and this is used as the reference phasor for the complete circuit. The voltage V_1 across R_1 is in phase with I and the applied voltage V is equal to the phasor sum of the voltages V_1 and V_2; this is shown by Fig. 1.28(c). ϕ is the phase angle between the applied voltage V and the supply current I.

Example 1.11

Determine the applied voltage V, the supply current I, and the phase angle between them if, in the circuit of Fig. 1.28(a), $R_1 = 47\,\Omega$, $R_2 = 100\,\Omega$, $C = 2.2\,\mu F$, the current in R_2 is 10 mA, and the frequency is $5000/2\pi$ Hz. Also calculate the power dissipated in the circuit.

Solution

$V_2 = 10 \times 10^{-3} \times 100 = 1\,\text{V}$.
$X_C = 90.9 \angle -90°\,\Omega$.
Hence $I_C = 1/90.9 \angle -90° = 11 \angle 90°\,\text{mA}$.
The total current is $I = \sqrt{(10^2 + 11^2)} \angle \tan^{-1}(-11/10)$
$= 14.87 \angle -47.7°\,\text{mA}$.   (*Ans.*)
The voltage V_1 is in phase with the total current and is equal to $14.87 \times 10^{-3} \times 47 = 0.7\,\text{V}$. The in-phase component of V_2 is equal to $1 \cos 47.7° = 0.673\,\text{V}$ and the quadrature component is $1 \sin 47.7° = 0.74\,\text{V}$. Hence, the applied voltage is
$$V = \sqrt{[(0.7 + 0.673)^2 + 0.74^2]} = 1.56\,\text{V}.   (\textit{Ans.})$$
The phase angle between the applied voltage and the supply current is
$$\phi = \tan^{-1}(0.74/1.373) = 28.3°.   (\textit{Ans.})$$
The power dissipated in the circuit is
$$(10 \times 10^{-3})^2 \times 100,\ \text{plus}\ (14.87 \times 10^{-3})^2 \times 47 = 10 + 10.39\,\text{mW}$$
$$= 20.39\,\text{mW}. (\textit{Ans.})$$

The resonant circuit

Resonance is said to have occurred in an $R,\,L,\,C$ circuit when the current supplied to the circuit is in phase with the applied voltage. A resonant circuit may consist of an inductor and a capacitor connected in either parallel or in series. Resonant circuits are widely employed in communication engineering because of their ability to be tuned to select a band of frequencies from a range of applied frequencies. The important parameters of a resonant circuit are its *resonant frequency*, its *Q factor*, and its *3 dB bandwidth*.

The series resonant circuit

A series resonant circuit consists of an inductor and a capacitor connected in series as shown by Fig. 1.29. The resistance R represents

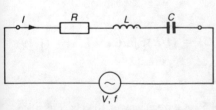

Fig 1.29 The series resonant circuit

the total resistance of the circuit, including the self-resistance r of the inductor and, perhaps, the series loss resistance of the capacitor. At any given frequency f the effective reactance X_T of the circuit is equal to the difference between the reactance X_L of the inductor and the reactance X_C of the capacitor. The magnitude of the circuit's impedance is

$$|Z| = \sqrt{[R^2 + (\omega L - 1/\omega C)^2]} \ \Omega \tag{1.15}$$

The impedance varies with change in frequency because of the $(\omega L - 1/\omega C)$ term and it has its minimum value when the effective reactance is equal to zero, when $|Z| = \sqrt{R^2} = R$ ohms. Since at this frequency the impedance of the circuit is purely resistive the current flowing in the circuit is in phase with the applied voltage. This is the condition for resonance. The resonant current I_0 is equal to V/R and it is the maximum value of the current that flows in the circuit.

At the resonant frequency f_0,

$$\omega_0 L = 1/\omega_0 C, \quad \omega_0^2 = 1/LC, \quad \omega_0 = 1/\sqrt{LC} \ \text{and}$$
$$f_0 = 1/(2\pi \sqrt{LC}) \ \text{Hz.} \tag{1.16}$$

Note that the resonant frequency is independent of the value of the resistance in the circuit.

Q factor

Although at the resonant frequency the effective reactance of the circuit is zero, the two reactances themselves are not equal to zero and voltages $V_L = I_0 \omega_0 L$, and $V_C = I_0/\omega_0 C$, are developed across the two components. Since $I_0 = V/R$, $V_L = V\omega_0 L/R$ and $V_C = V/\omega_0 CR$, and R is (usually) very much smaller than either $\omega_0 L$ and $1/\omega_0 C$, the voltages V_L and V_C may be much larger than the applied voltage V. This effect is known as the *voltage magnification* of the circuit. Voltage magnification is commonly employed in radio circuits but it is avoided in power circuits because it may result in a dangerously high voltage appearing across the capacitor.

The *Q factor* of a series-resonant circuit is the ratio (capacitor voltage)/(applied voltage). Therefore,

$$Q = V_C/V = 1/\omega_0 CR = \omega_0 L/R \tag{1.17}$$

The latter relationship is also commonly employed to indicate the quality of an inductor at any frequency f, i.e.

$$Q = \omega L/r \tag{1.18}$$

From equation (1.17),

$$Q = \omega_0 L/R = (1/R)(L/\sqrt{LC}) = (1/R)(\sqrt{L/C}) \tag{1.19}$$

Example 1.12

A 20 mH inductor has a Q factor of 50 at a frequency of 10 kHz. Calculate its self-resistance.

Solution
$50 = (2\pi \times 10^4 \times 20 \times 10^{-3})/r$
or $r = 25.1\ \Omega$.   (*Ans.*)

Example 1.13

A circuit consists of a 5 mH inductor with a self-resistance of 10 Ω connected in series with a 47 nF capacitor. Calculate (*a*) the resonant frequency, (*b*) the Q factor of (i) the inductor and (ii) the circuit, and (*c*) the voltage that appears across the capacitor when 1 V at the resonant frequency is applied to the circuit.

Solution
(*a*) $f_0 = 1/(2\pi \sqrt{(5 \times 10^{-3} \times 47 \times 10^{-9})}) = 10.382\ \text{kHz}$.   (*Ans.*)
(*b*) Q of inductor = Q of circuit =
    $(2\pi \times 10.382 \times 10^3 \times 5 \times 10^{-3})/10 = 32.6$.   (*Ans.*)
(*c*) $V_C = QV = 32.6 \times 1 = 32.6\ \text{V}$.   (*Ans.*)

Selectivity of a series-resonant circuit

The selectivity of a series-resonant circuit is its ability to discriminate between signals at different frequencies. If a constant-voltage source of variable frequency is applied to a series-resonant circuit the current that flows in the circuit will vary with frequency in the manner shown by Fig. 1.30. At low frequencies the effective reactance of the circuit is very high, because X_C is large and X_L is small, and so only a small current is able to flow. With increase in frequency the reactance of the capacitor falls and the reactance of the inductor increases and so the effective capacitive reactance, $X_C - X_L$, decreases. At the resonant frequency the effective reactance is zero, because $X_C = X_L$, and then the maximum current I_0 flows in the circuit. When the

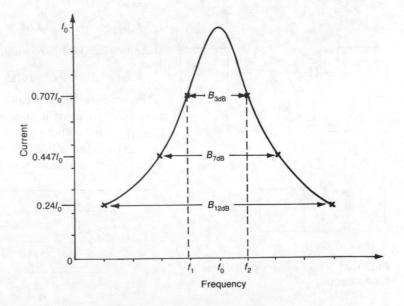

Fig. 1.30 Selectivity of a series-tuned circuit

frequency is increased above the resonant frequency the inductive reactance becomes larger than the capacitive reactance and so the effective reactance becomes inductive and increases with further increase in frequency. As a result the current now decreases with increase in the frequency.

The selectivity of a resonant circuit is often expressed in terms of its *3 dB bandwidth*. This is the band of frequencies over which the current is not less than 0.707 times the resonant current I_0.

The shape of the current/frequency curve depends upon the Q factor of the circuit. When the bandwidth B is equal to f_0/Q the current is equal to $0.707I_0$, when $B = 2f_0/Q$, $I = 0.447I_0$, and when $B = 4f_0/Q$, $I = 0.24I_0$. These values correspond to -3, -7 and -12 dB respectively. The higher the Q factor, i.e the lower the circuit resistance, the more selective will be the tuned circuit.

At the two 3 dB frequencies, f_1 and f_2, $I_1 = I_2 = I_0/\sqrt{2}$. Since the resistance R of the circuit is constant at these frequencies $X_T = R$. The impedance of the circuit is then

$$Z = \sqrt{(R^2 + X_T^2)} = \sqrt{(R^2 + R^2)} = \sqrt{2}R \quad \text{and} \quad I = V/\sqrt{2}R = I_0/\sqrt{2}$$

At frequency f_2,

$$\omega_2 L - 1/\omega_2 C = R$$

$$\omega_2^2 L - 1/C = \omega_2 R \tag{1.20}$$

At frequency f_1,

$$-\omega_1^2 L + 1/C = \omega_1 R \tag{1.21}$$

Adding equations (1.20) and (1.21) gives

$$L(\omega_2^2 - \omega_1^2) = R(\omega_1 + \omega_2)$$

$$L(\omega_2 - \omega_1) = R^\dagger, \quad (\omega_2 - \omega_1)/\omega_0 = R/\omega_0 L$$

$$f_0/(f_2 - f_1) = \omega_0 L/R = Q$$

Therefore the 3 dB bandwidth is

$$B_{3\text{dB}} = f_2 - f_1 = f_0/Q \tag{1.22}$$

The series-resonant circuit can be used to select a wanted band of frequencies by using it to connect two circuits together, as in Fig. 1.31(*a*), or the output signal can be taken from the capacitor, as in Fig. 1.31(*b*). The voltage across the capacitor varies with frequency in very nearly the same way as the current in the circuit.

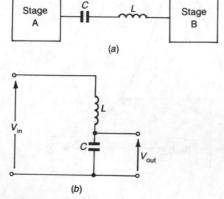

(a)

(b)

Fig. 1.31 Two uses for a series-tuned circuit

$\dagger$

$$\omega_1 + \omega_2 \sqrt{\dfrac{\omega_2 - \omega_1}{\omega_2^2 - \omega_1^2}}$$

$$\omega_2^2 + \omega_2\omega_1$$

$$\underline{\quad - \quad \omega_1^2 - \omega_2\omega_1}$$

$$- \quad \omega_1^2 - \omega_2\omega_1$$

$$\underline{\quad - \qquad - \quad}$$

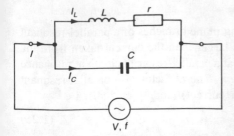

Fig. 1.32 The parallel-tuned circuit

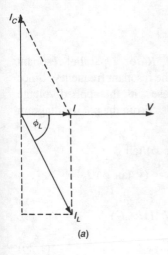

(a)

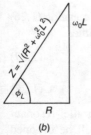

(b)

Fig. 1.33 (a) Phasor diagram; (b) impedance triangle, of a parallel-tuned circuit

The parallel-resonant circuit

A parallel-resonant circuit consists of an inductor and a resistor connected in parallel as shown by Fig. 1.32. The inevitable self-resistance of the inductor is represented by the resistor r. At low frequencies the current in the inductive branch is larger than the capacitor current, since $\omega L < 1/\omega C$, but at high frequencies the opposite is true. At some frequency in between the two currents are very nearly equal and their resultant — which is the supply current I — is in phase with the applied voltage. This in the condition for parallel resonance. At the resonant frequency the supply current is small and this means that the impedance of the circuit at resonance, known as the dynamic impedance or resistance, is very large.

Resonant frequency

The current I_L in the inductive branch of the circuit is

$$I_L = V/\sqrt{(R^2 + \omega_0^2 L^2)}$$

and lags the applied voltage by angle $\phi = \tan^{-1}(\omega_0 L/R)$. The capacitor current is $I_C = V\omega_0 C$ and this leads the applied voltage by $90°$. The phasor diagram for the circuit is shown in Fig. 1.33(a); the supply current is the phasor sum of I_C and I_L and it is in phase with the applied voltage. The quadrature (reactive) component of I_L is $I_L \sin \phi$ and this is in anti-phase with I_C. The in-phase component of I_L is $I_L \cos \phi$ and it is in phase with the applied voltage V. From the phasor diagram, $I_C = I_L \sin \phi$.

From the impedance triangle of the inductive branch (see Fig. 1.33(b)),

$$\sin \phi_L = \omega_0 L/\sqrt{(R^2 + \omega_0^2 L^2)}$$

Hence

$$V\omega_0 C = V/\sqrt{(R^2 + \omega_0^2 L^2)} \times \omega_0 L/\sqrt{(R^2 + \omega_0^2 L^2)}$$
$$= V\omega_0 L/\sqrt{(R^2 + \omega_0^2 L^2)}.$$

Therefore

$$R^2 + \omega_0^2 L^2 = L/C \quad \text{and} \quad \omega_0^2 = 1/LC - R^2/L^2$$

Therefore the resonant frequency f_0 is given by

$$f_0 = 1/[2\pi \sqrt{(1/LC - R^2/L^2)}] \text{ Hz} \tag{1.23}$$

Very often the second term R^2/L^2 is very much smaller than the first term $1/LC$ and it may then be neglected with very little error being introduced. If this is done,

$$f_0 = 1/2\pi \sqrt{LC} \text{ Hz} \tag{1.24}$$

which is the same as the expression for the resonant frequency of a series-resonant circuit.

Q factor

The currents I_C and I_L flowing in the branches of a parallel-resonant circuit may be several times larger than the current taken from the voltage source. This means that a parallel-resonant circuit at resonance provides *current magnification*. The Q factor of a parallel-resonant circuit is the ratio (capacitor current)/(supply current), i.e.

$$Q = I_C/I \tag{1.25}$$

The Q factor of a parallel-resonant circuit is equal to the Q factor of the inductor, i.e $\omega_0 L/r$, as long as the capacitor is loss free.

Dynamic impedance

The *dynamic impedance*, or *resistance*, R_d of a parallel-resonant circuit is the impedance of that circuit at the resonant frequency. Since at resonance the supply current is in phase with the applied voltage the dynamic impedance is purely resistive. From the phasor diagram shown in Fig. 1.33(a)

$$\tan \phi = (I_L \sin \phi)/I \qquad I = (I_L \sin \phi)/\tan \phi$$

$$R_d = V/I = (V \tan \phi)/I_L \sin \phi = (V \tan \phi)/I_C = (V \tan \phi)/V\omega_0 C$$

$$= \tan \phi/\omega_0 C = (\omega_0 L/r)(1/\omega_0 C)$$

or

$$R_d = L/Cr \ \Omega \tag{1.26}$$

Multiplying both the numerator and the denominator of equation (1.26) by ω_0 gives $\omega_0 L/\omega_0 Cr$ and this may be written as either

$$R_d = Q\omega_0 L \ \Omega \tag{1.27}$$

or as

$$R_d = Q/\omega_0 C \ \Omega \tag{1.28}$$

It is now possible to confirm equation (1.25). If a parallel-resonant circuit is supplied with a current I a voltage IR_d will be dropped across it. The current I_C in the capacitor is

$$I_C = V/X_C$$

$$= IR_d\omega_0 C = I\omega_0 LC/Cr$$

$$= I\omega_0 L/r = QI$$

The current in the inductive branch is

$$I_L = V/\sqrt{(r^2 + \omega_0^2 L^2)}$$

$$= V/[\omega_0 L \sqrt{(1 + R^2/\omega_0^2 L^2)}]$$

$$= [(L/Cr)I]/[\omega_0 L \sqrt{(1 + 1/Q^2)}]$$

$$= IL/CR\omega_0 L \sqrt{[1 + 1/Q^2)}]$$

or

$$I_L = QI/\sqrt{(1 + 1/Q^2)} \ \text{A} \tag{1.29}$$

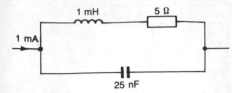

Fig. 1.34

If Q is larger than about 10 the $1/Q^2$ term is negligibly small and then $I_L = I_C = QI$.

Example 1.14

For the circuit shown in Fig. 1.34 calculate (*a*) the resonant frequency, (*b*) the Q factor, (*c*) the dynamic resistance , and (*d*) the capacitor current when 1 mA at the resonant frequency is supplied to the circuit.

Solution

(*a*) $f_0 = (1/2\pi)/\sqrt{(1/(25 \times 10^{-12}) - 25/(1 \times 10^{-6})]}$
   $= 31.821$ kHz.   (*Ans.*)

Note that if the approximate expression for the resonant frequency is used then
   $f_0 = 1/[2\pi \sqrt{(25 \times 10^{-12})} = 31.831$ kHz.   (*Ans.*)

(*b*) $Q = \omega_0 L/r = (2\pi \times 31.831 \times 10^3 \times 1 \times 10^{-3})/5 = 40 \, \Omega$.   (*Ans.*)

(*c*) $R_d = L/Cr = (1 \times 10^{-3})/(25 \times 10^{-9} \times 5) = 8000 \, \Omega$.   (*Ans.*)

(*d*) $I_C = QI = 40$ mA.   (*Ans.*)

Selectivity

The impedance of a parallel-resonant circuit will vary with change in frequency in exactly the same way as the variation of current with frequency in a series-resonant circuit. Thus, Fig. 1.30 will show the impedance/frequency characteristic of a parallel-resonant circuit if the vertical axis is changed to read magnitude of impedance instead of current. The maximum impedance, which is the dynamic resistance R_d, occurs at the resonant frequency and the selectivity is expressed in terms of the 3 dB bandwidth. As for the series circuit,

$$B_{3dB} = f_0/Q \qquad\qquad (1.30)$$

Figure 1.35 shows one way in which a parallel-resonant circuit may be connected in the collector circuit of a transistor to provide the necessary selectivity for an RF amplifier.

Example 1.15

A parallel-resonant circuit has a capacitor of 100 pF and it is resonant at 465 kHz. If the 3 dB bandwidth of the circuit is 10 kHz calculate (*a*) the inductance and resistance of the inductor and (*b*) the dynamic impedance of the circuit.

Solution

(*a*) $Q = f_0/B_{3dB} = (465 \times 10^3)/(10 \times 10^3) = 46.5$.
This figure is also the Q factor of the inductor and hence
   $46.5 = \omega_0 L/r = 1/\omega_0 Cr$.
Therefore, $r = 1/(2\pi \times 465 \times 10^3 \times 100 \times 10^{-12} \times 46.5) = 73.61 \, \Omega$.   (*Ans.*)
$L = 1/[4\pi^2 \times (465 \times 10^3)^2 \times 100 \times 10^{-12}] = 1.17$ mH.   (*Ans.*)

(*b*) $R_d = L/Cr = (1.17 \times 10^{-3})/(7361 \times 10^{-12}) = 158.95$ kΩ.   (*Ans.*)

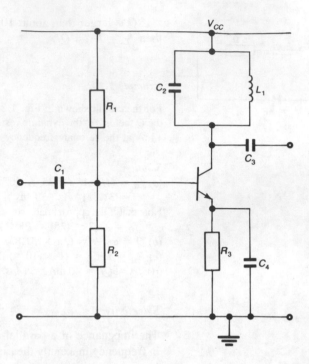

Fig. 1.35 Tuned radio-frequency amplifier

Alternatively

$$R_d = Q\omega_0 L = 46.5 \times 2\pi \times 465 \times 10^3 \times 1.17 \times 10^{-3} = 158.95 \text{ k}\Omega.$$

Power factor and power factor correction

The instantaneous power dissipated in an a.c. circuit is equal to the product of the instantaneous values of the current flowing in the circuit and the voltage across the circuit. The mean power dissipated is the product of the r.m.s. values of the current, the voltage and the power factor, i.e. $P = VI \times$ power factor.

The power factor is defined as

$$\text{Power factor} = (\text{true power})/(\text{apparent power}) \qquad (1.31)$$

If both the current and the voltage are of sinusoidal waveform the power factor is equal to $\cos\phi$, where ϕ is the phase angle between the current and the voltage. Then

$$P = VI \cos\phi \text{ W} \qquad (1.32)$$

When the power factor is said to be lagging, or leading, it means that the current is lagging, or leading, the voltage. Thus, a lagging power factor refers to an inductive circuit, and a leading power factor refers to a capacitive circuit.

The *apparent power S* is the term given to the product of the circuit current and the applied voltage, i.e $S = VI$ volt-amps. The reactive volt-amps are equal to $VI \sin\phi$ var.

Example 1.16

A resistor and a capacitor are connected in series and take 150 W power at a power factor of 0.8 leading from a 240 V, 50 Hz supply. Calculate (*a*) the current that flows in the circuit and (*b*) the value of each component.

Solution

(*a*) $0.8 = 150/(240I)$. $I = 150/(240 \times 0.8) = 0.78$ A.   (*Ans.*)

(*b*) $\phi = \cos^{-1}0.8 = 36.9°$. $|Z| = V/I = 240/0.78 = 307.7\,\Omega$. Therefore, $R = 307.7 \cos 36.9° = 246\,\Omega$.   (*Ans.*)

$X_C = 307.7 \sin 36.9° = 184.8\,\Omega = 1/100\pi C$, and

$\qquad C = 17.2\ \mu\text{F}$.   (*Ans.*)

Example 1.17

A coil of inductance L and self-resistance r is connected across a 240 V, 50 Hz supply. The current in the coil is 0.2 A and the power dissipated in the coil is 5 W. Calculate the values of (*a*) L and r, (*b*) the power factor, (*c*) the apparent power and (*d*) the reactive volt-amps.

Solution

(*a*) $P = |I|^2 r$, $r = 5/0.2^2 = 125\,\Omega$.   (*Ans.*)

$Z = V/I = 240/0.2 = 1200\,\Omega$. $X_L = \sqrt{(1200^2 - 125^2)} = 1193\,\Omega$.

$L = 1193/100\pi = 3.8$ H.   (*Ans.*)

(*b*) $\phi = \tan^{-1}(1193/125) = 84°$. Power factor $= \cos 84° = 0.1$.   (*Ans.*)

(*c*) Apparent power $S = 240 \times 0.2 = 48$ VA.   (*Ans.*)

(*d*) Reactive volt-amps $= 240 \times 0.2 \sin 84° = 47.74$ var.   (*Ans.*)

Power factor correction

When an electrical machine, or other equipment, is rated at a certain current and voltage it means that these are the highest values that the machine can handle for a lengthy period of time without damage. The phase difference ϕ between the supply voltage and the current taken by the machine is determined by the nature of the load presented by that machine. If the power factor is unity ($\phi = 0$), the volt-amp rating of the machine will also be the power that the machine can dissipate. But if the power factor is less than unity, 0.75 say, the power that the machine is able to dissipate will be only 0.75 times the volt-amp rating. Since the user is charged by the electricity company for the volt-amps delivered this represents a waste of money for the user. A non-unity power factor also means that the conductors that convey the supply to the user's premises will be conducting a higher current to supply a given power than if the power factor were unity.

Consider, as an example, a load that when connected to the 240 V mains power supply consumes a power of 2000 W at a power factor of 0.8. The current taken from the supply is $I = 2000/(240 \times 0.8) = 10.42$ A. If the power factor were to be improved to 0.94 the current taken from the supply would fall to $I = 2000/(240 \times 0.94) = 8.87$ A, and if the power could be improved to unity the current would be only $2000/240 = 8.33$ A. Thus, improving the power factor reduces the current taken by a load for the same power dissipation. This reduces the I^2R power losses in lines and conductors and increases the overall efficiency of the system.

Example 1.18

A 120 kVA rated cable is used to supply a load. Determine whether the load may dissipate 100 kW power if the power factor is (a) 0.9 and (b) 0.8.

Solution
(a) The maximum power that may be dissipated $= 120 \times 0.9 = 108$ kW. Hence 100 kW power may be dissipated.   (*Ans.*)
[b] Maximum power $= 120 \times 0.8 = 96$ kW. Now the 100 kW power cannot be dissipated.   (*Ans.*)

Power factor correction

The nearer the power factor of a load is to unity the larger will be the power that the load can dissipate for a given volt-amp rated system. Most of the loads used in heavy current systems are inductive, such as large electrical motors, and, electrically, can be considered to consist of an inductor in series with a resistor. Power factor correction can be achieved by connecting a suitable value capacitor in parallel with the load. The basic idea is illustrated by Fig. 1.36(a). Before the power factor correction capacitor is fitted, the phasor diagram of the load is as shown by Fig. 1.36(b). The power factor of the load is $\cos \phi_L$. When the power factor correction capacitor is connected in parallel with the load the phasor diagram is altered to that shown in Fig. 1.36(c). The current I taken from the supply is the phasor sum of I_C and I_L. The phase angle ϕ between the supply current I and the supply voltage V is smaller than the angle ϕ_L and so $\cos \phi$ is nearer than $\cos \phi_L$ to the desired ideal value of unity. It can be seen that the in-phase or active component of the current has not been altered. The greater the amount of power factor correction wanted, i.e. the smaller the required angle ϕ, the larger will be the necessary value of the correction capacitor. Very large capacitance values prove to be very expensive in high-power systems and so most power factor correction arrangements are employed to give an increase in the power factor to a figure less than unity.

Example 1.19

A load takes a current of 30 A from a 240 V 50 Hz supply with a power factor of 0.68 lagging. Determine the component that could be connected in parallel with the load to bring the power factor to (a) unity and (b) 0.9.

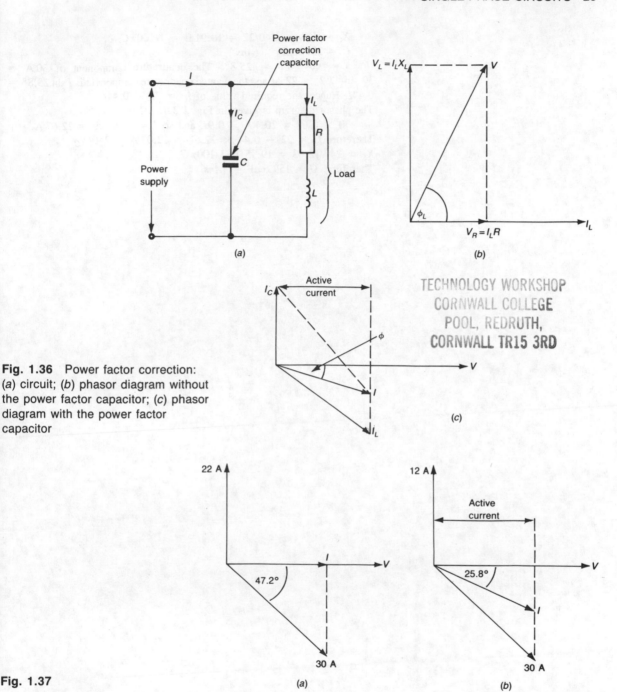

Fig. 1.36 Power factor correction: (*a*) circuit; (*b*) phasor diagram without the power factor capacitor; (*c*) phasor diagram with the power factor capacitor

Fig. 1.37

Solution

(*a*) Since the power factor is lagging the load is inductive and hence the parallel component must be a capacitor. $\phi_L = \cos^{-1} 0.68 = 47.2°$. From the phasor diagram shown in Fig. 1.37(*a*) it is required that

$$I_C = 30 \sin 47.2° = 22 \text{ A}.$$

$X_C = V/I_C = 240/22 = 10.91 \, \Omega = 1/100\pi C, \; or$
$C = 291.8 \, \mu\text{F}.$ (*Ans.*)

(*b*) $\phi = \cos^{-1}0.9 = 25.8°$. The quadrature component of 30 A $=$ 30 sin 47.2° $= 22$ A, and the wanted quadrature component is $I \sin 25.8° = 0.44I$. Hence, the required value of $I_C = 22 - 0.44I$.

The phasor diagram is given in Fig. 1.37(*b*). Now

$30 \times 0.68 = 20.4 \, \text{A} = 0.9I$, and so $I = 20.4/0.9 = 22.67 \, \text{A}$.

Therefore $I_C = 22 - 0.44 \times 22.67 = 12.03 \, \text{A} = 240/X_C$.

$X_C = 240/12.03 = 19.95 = 1/100\pi C$.

Therefore, $C = 159.6 \, \mu\text{F}.$ (*Ans.*)

2 Three-phase circuits

A three-phase supply consists of three sinusoidal voltages at the same frequency, usually 50 Hz, having the same amplitude, but with a mutual phase difference of 120°. The three phases are identified by being labelled as the red, yellow and blue phases. The three waveforms are shown in Fig. 2.1. It can be seen that the red voltage is the first to reach its peak positive value, followed by the yellow voltage one-third of a cycle later, and then by the blue voltage another one-third cycle later. The three voltages are known as the *phase voltages* and the order in which they reach their positive peak values is known as the *phase sequence*. There are two possible sequences through which the phase voltages may pass: the positive sequence, shown in Fig. 2.1, is red, yellow and blue, and the negative sequence is red, blue and yellow. The two phase sequences are shown by the phasor diagrams given in Figs 2.2(*a*) and (*b*). The positive phase sequence is very much the more commonly employed and its use will be assumed throughout this chapter.

The three-phase voltages are generated by a three-phase generator, or alternator, which basically has three identical coils mounted in a uniform magnetic field, as shown by Fig. 2.3. If the coils rotate within the magnetic field, or the coils are stationary and the magnetic field rotates around them, with an angular velocity of ω rads/s, an e.m.f. of E volts will be induced into each coil. Because of the physical positioning of the coils the induced e.m.f.s will be mutually 120° out of phase with one another. Normally the red phase is taken to be the reference phase and so the instantaneous voltages of the three phases are

$$e_R = E \sin \omega t \tag{2.1}$$

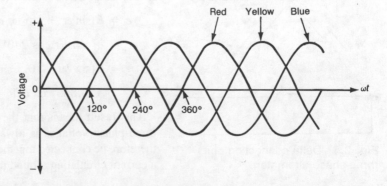

Fig. 2.1 Three-phase waveforms

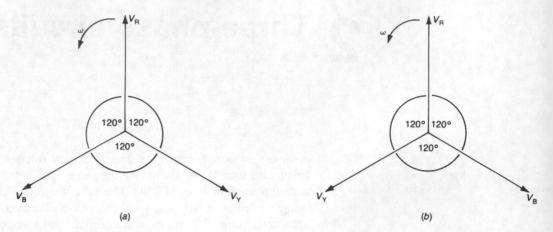

(a) *(b)*

Fig. 2.2 Phase sequences: *(a)* positive; *(b)* negative

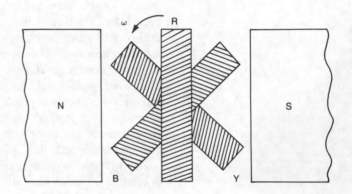

Fig. 2.3 Basic three-phase alternator

$$e_Y = E \sin (\omega t - 120°) \tag{2.2}$$

$$= E \sin (\omega t - 240°) = E \sin (\omega t + 120°) \tag{2.3}$$

The sum of the three instantaneous phase voltages is

$$e = e_R + e_Y + e_B$$

$$= E[\sin \omega t + \sin (\omega t - 120°) + \sin (\omega t - 240°)] \tag{2.4}$$

A trigonometric identity is $\sin (A - B) = \sin A \cos B - \sin B \cos A$ and using this equation (2.4) can be written as

$$e = E[\sin\omega t + \sin \omega t \cos 120° - \sin 120° \cos \omega t$$

$$+ \sin \omega t \cos 240° - \sin 240° \cos \omega t]$$

$$= E[\sin \omega t - 0.5 \sin \omega t - 0.866 \cos \omega t - 0.5 \sin \omega t$$

$$+ 0.866 \cos \omega t] = 0$$

This result means that the sum of the instantaneous values of the three phase voltages is always equal to zero. The three coils may therefore be connected together in delta, as shown by Fig. 2.4, without a current circulating around the loop. In practice, the delta connection

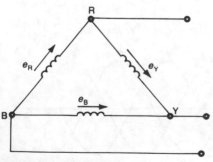

Fig. 2.4 Delta connection of a three-phase alternator

is not often used for a three-phase generator, or alternator. Instead, the three windings are usually *star-connected* as shown by Fig. 2.5(*a*). If the star-connected generator is connected via four conductors to a star-connected load, as in Fig. 2.5(*b*), currents I_R, I_Y and I_B will flow in the red, yellow and blue lines respectively. The sum of the line currents flows via the neutral line to the neutral, or *star*, point of the generator and this point is usually connected to earth. If the load is balanced, i.e. $Z_R = Z_Y = Z_B$, the three currents will be equal to one another and mutually 120° out of phase. Hence, following the previous analysis the sum of the line currents is equal to zero. This means that a *balanced* system does not require the neutral conductor and so a three-wire star-connected system can be employed. This is shown by Fig. 2.6.

Delta-connected loads are also commonly employed and Fig. 2.7 shows a delta-connected load supplied by a star-connected generator.

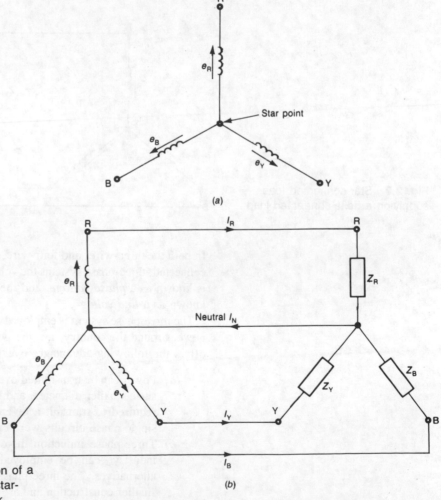

Fig. 2.5 (*a*) Star connection of a three-phase alternator; (*b*) star-connected four-wire network

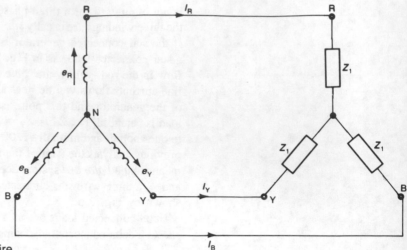

Fig. 2.6 Star-connected three-wire network

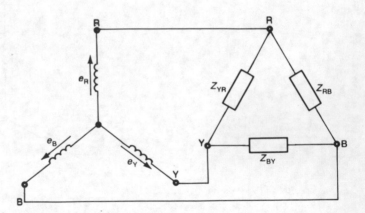

Fig. 2.7 Star-connected load supplying a delta-connected load

In both the three-wire, and four-wire, star-connected, and the delta-connected, three-phase systems the voltage between a line and neutral is known as a *phase voltage*, and the voltage between two lines is known as a *line voltage*.

The three-phase system is employed for the distribution of electrical power around the country, i.e. the public mains supply, because it offers the following advantages over alternative methods:

(a) Power can be transmitted over cables with greater economy since smaller diameter, and hence cheaper, conductors are required to transmit a given power than if three separate single-phase circuits were employed.

(b) Three-phase induction motors, which are widely used in industry, can be employed instead of the single-phase alternatives. The three-phase machines are of simpler and smaller construction and are more robust than single-phase

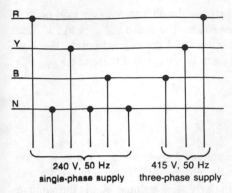

Fig. 2.8 Mains supply system

machines. In addition, they are self-starting whereas single-phase machines must be provided with starting circuitry.

(*c*) Two voltages are made available; these are the phase voltage and the line voltage. The public electricity supply in the UK is delivered from the power stations to the local neighbourhood using the four-wire three-phase system. The electricity supply to houses, shops and offices is taken from between one line and neutral and delivers the phase voltage of 240 V. The current flows along the line conductor to the domestic load and then returns to the star point of the generator via the neutral line. Although the division of the total domestic load into three parts is made as equal as possible inevitable fluctuations in the demand ensure that the total load is always unbalanced and this is why the fourth, neutral, wire is necessary. Industrial loads are usually balanced and they are supplied by the three-wire system. Small factories, etc. that employ three-phase machinery which take a power of more than about 20 kW are supplied with the 415 V line voltage. The arrangement is shown by Fig. 2.8.

The electrical generators, known as alternators, used to generate the three-phase voltages usually have their coils connected in star, but three-phase transformers, electric motors and other high-power loads may be either delta, or star, connected.

The star connection

Figure 2.9 shows a balanced three-wire star-connected three-phase system with all the line currents and line/phase voltages labelled. The

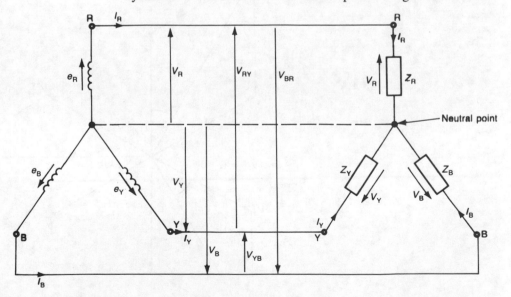

Fig. 2.9 Balanced three-wire star-connected three-phase system

line currents I_R, I_Y and I_B flow from the alternator to the load neutral point via their respective phase loads. The phase voltages V_R, V_Y and V_B are the voltages between a line and the neutral point. The line voltages are the voltages between two lines and these are V_{RY}, V_{YB} and V_{BR}.

Line voltages

The line voltage V_{RY} is the voltage between the red line and the yellow line and it is equal to the phasor difference between V_R and V_Y, i.e. $V_{RY} = V_R - V_Y$. Similarly, line voltage V_{YB} is the voltage between the yellow line and the blue line and it is equal to $V_Y - V_B$. Lastly, $V_{BR} = V_B - V_R$. The three phase voltages and the three line voltages are shown by the phasor diagram given in Fig. 2.10. The phasor difference between two phasors, say V_R and V_Y, is obtained by reversing the direction of V_Y and then adding it to V_R. It can be seen that the line voltages are, just like the phase voltages, mutually 120° out of phase with one another. Each line voltage leads a phase voltage by 30°, e.g. V_{RY} leads V_R by 30°.

Figure 2.11 shows the part of the phasor diagram associated with line voltage V_{RY}. From this diagram, $\cos 30° = AB/V_R$, and so $AB = V_R \cos 30°$. Hence

$$V_{RY} = 2AB = 2V_R \cos 30° = \sqrt{(3)}V_R$$

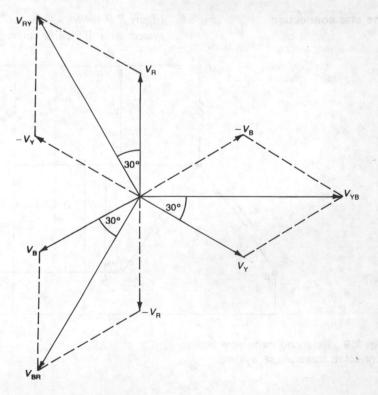

Fig. 2.10 Phasor diagram of the voltages in a balanced star-connected network

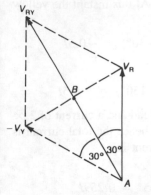

Fig. 2.11 Part of the phasor diagram in Fig. 2.10.

A similar result can also be obtained by considering either of the other two line voltages and this means that the amplitude of a line voltage is equal to $\sqrt{3}$ times the amplitude of a phase voltage, i.e.

$$V_L = \sqrt{(3)}V_{ph} \tag{2.5}$$

In the UK electricity supply system the line voltage is 415 V and the phase voltage is $415/\sqrt{3} = 239.6$ V, which is always taken to be 240 V.

Line currents

The phase current I_{ph} flowing in each phase of the load is equal to the phase voltage divided by the phase impedance, i.e. $I_{ph} = V_{ph}/Z_{ph}$. For a balanced load $I_{ph} = I_R = I_Y = I_B$. The phase of each line current relative to the phase voltage is determined by the phase angle $\phi = \tan^{-1}(X/R)$ of the phase impedance. In a star-connected load the phase currents are equal to the line currents, this should be clear from an inspection of Fig. 2.9.

Figure 2.12 gives the phasor diagram for a balanced system in which $\phi = 20°$ lagging. The sum of the three line currents is equal to zero since they are of equal amplitude and are mutually 120° out of phase (see equation (2.4) again). However, at any instant in time there is a current flowing into the load from the source in one, or perhaps two, of the conductor(s), and a current flowing out of the load towards the source in the other two, or perhaps one, conductor(s). This is, of course, because the three currents each vary in the same way as the voltages shown in Fig. 2.1. Consider, for example, the instant

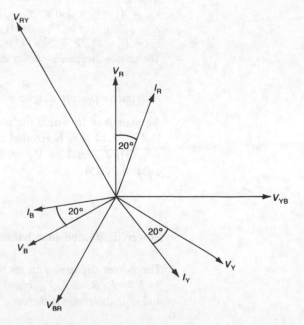

Fig 2.12 Phasor diagram of a balanced star-connected system with a phase angle of 20° lagging

when the red current $i_R = I \sin 90° = I$. At this instant the yellow current will then be

$$i_Y = I \sin (90° - 120°) = I \sin (-30°) = -0.5I$$

and the blue current will be

$$i_B = I \sin (90° - 240°) = I \sin (-150°) = -0.5I$$

There is now a current I flowing into the load and a current of $2 \times 0.5I = I$ flowing away from the load and hence the total current is zero. At a later instant when the red current is

$$i_R = I \sin 135° = 0.707I,$$
$$i_Y = I \sin (135° - 120°) = I \sin 15° = 0.259I$$

and

$$i_B = I \sin (135° - 240°) = I \sin (-105°) = -0.966I$$

There is now a total current of $(0.707 + 0.259) = 0.966I$ flowing into the load and $0.996I$ flowing out of the load so that once again the total current is zero.

If the load is unbalanced then the amplitudes and/or the phase angles of the three phase (=line) currents will differ from one another. The line currents will then have a non-zero sum and a fourth, neutral, conductor will be required to carry the neutral current.

Example 2.1

Each phase of a balanced three-phase load consists of a 100 Ω resistor in series with a 22 μF capacitor. If the supply voltage is 415 V at 50 Hz calculate the line currents and draw the phasor diagram for the system.

Solution

$V_{ph} = 414/\sqrt{3} = 240$ V. $X_C = 1/(100\pi \times 22 \times 10^{-6}) = 144.7$ Ω.
$Z = \sqrt{(100^2 + 144.7^2)} \angle \tan^{-1}(-144.7/100) = 175.9 \angle -55.4°$ Ω.
$I_{ph} = I_L = V_{ph}/Z = 240/175.9 \angle -55.4° = 1.365 \angle 55.4°$ A.   (*Ans.*)
The phasor diagram is shown in Fig. 2.13.

Alternative phasor diagram

Another way in which the phasor diagram can be drawn is shown by Fig. 2.14. The horizontal component of V_Y is equal to $V_Y \cos 30° = \sqrt{(3)}V_Y/2$ and so $V_Y = 2\sqrt{(3)}V_Y/2 = \sqrt{(3)}V_Y$. This confirms equation (2.5).

Power dissipated in a balanced three-phase load

The power dissipated in each phase of a three-phase load is given by $P = I_{ph}^2 R$, or $I_{ph}V_{ph} \cos \phi$, where R is the resistance of the load and ϕ is the phase difference between I_{ph} and V_{ph}. The total power

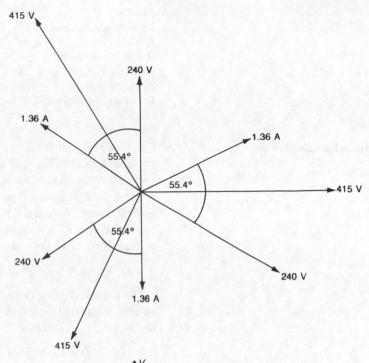

Fig. 2.13

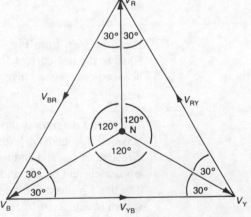

Fig. 2.14 Alternative phasor diagram for a balanced star-connected load

dissipated in the load is the sum of the powers dissipated in each phase. For a balanced system each phase of the load dissipates the same power and so this must be equal to one-third of the total power. Therefore

$$P = 3V_{ph}I_{ph} \cos \phi = 3(V_L/\sqrt{3}) I_L \cos \phi$$

or

$$P = \sqrt{(3)}V_LI_L \cos \phi \quad W \tag{2.6}$$

Example 2.2

Three inductors, each having an inductance of 50 mH and 10 Ω resistance, are star-connected to a 415 V 50 Hz three-phase supply. Calculate (*a*) the power factor of the load, and (*b*) the total power dissipated.

Solution

(a) For each inductor, $X_L = 100\pi \times 50 \times 10^{-3} = 15.71\,\Omega$.

$Z = \sqrt{(10^2 + 15.71^2)} \angle \tan^{-1}(15.71/10) = 18.62 \angle 57.5°\,\Omega$.

Hence, power factor $= \cos 57.5° = 0.537$.   (*Ans.*)

(b) Phase voltage $V_{ph} = 240\,V$. $I_{ph} = 240/18.62 = 12.89\,A$.

Power in each phase $= I_{ph}^2 R = 12.89^2 \times 10 = 1661.5\,W$.

Therefore, the total power dissipated $= 3 \times 1661.5 = 4984.5\,W$.   (*Ans.*)

The delta connection

Figure 2.15 shows a balanced load connected in the delta configuration. The line voltage V_{RY} between the red and the yellow lines is directly applied across the impedance connected between the R and Y load terminals. Similarly, the line voltages V_{YB} and V_{BR} are applied directly across the phase impedances between the Y and B, and the B and R, terminals respectively. This means that the line voltage is equal to the phase voltage, i.e.

$$V_L = V_{ph} \tag{2.7}$$

The current flowing in each phase impedance, known as the phase current, is equal to the phase voltage divided by the phase impedance, i.e. $I_{ph} = V_{ph}/Z_{ph}$. Thus

$$I_{RY} = V_{RY}/Z_1 = V/Z_1 \qquad I_{YB} = V_{YB}/Z_1 = V\angle -120°/Z_1$$

and

$$I_{BR} = V_{BR}/Z_1 = V\angle -240°/Z_1$$

It can be seen from Fig. 2.15 that now the phase currents are not equal to the line currents but are each equal to the phasor difference between two phase currents. Referring to Fig. 2.15,

$$I_R = I_{RY} - I_{BR} \qquad I_Y = I_{YB} - I_{RY} \quad \text{and} \quad I_B = I_{BR} - I_{YB}$$

The phasor diagram is shown in Fig. 2.16(a). The phasor difference between I_{RY} and I_{BR}, for example, is obtained by reversing the direction of I_{BR} and adding it to I_{RY} to get I_R. Each line current lags a phase current by 30° and the line currents are mutually 120° apart. The line (= phase) voltages have been added to the phasor diagram shown in Fig. 2.16(b) assuming that, in each phase, the current lags the voltage by angle ϕ.

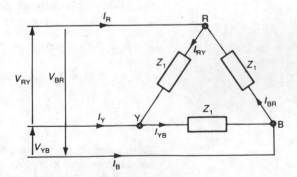

Fig. 2.15   Delta-connected balanced load

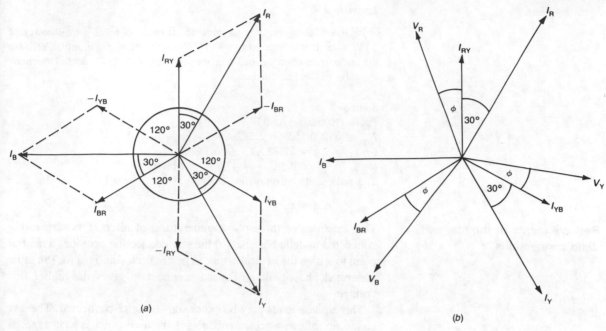

Fig. 2.16 (a) Phasor diagram of the currents in a balanced delta-connected network; (b) the same diagram with added line voltages

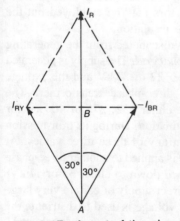

Fig. 2.17 A part of the phasor diagram shown in Fig. 2.16(a)

To determine the relationship between the line and phase currents in a delta-connected system the phasor diagram given in Fig. 2.17 can be obtained from Fig. 2.16. From this phasor diagram,

$$I_R = 2AB = 2I_{RY} \cos 30° = \sqrt{(3)}I_{RY}$$

Therefore, for a delta-connected load the line current is $\sqrt{3}$ times the phase current:

$$I_L = \sqrt{(3)}I_{ph}. \tag{2.8}$$

Example 2.3

The three inductors of example 2.2 are now connected in delta across the same three-phase supply. Calculate the total power dissipated. What is the ratio (delta power)/(star power)?

Solution
Now $V_{ph} = V_L = 415$ V, and $I_{ph} = 415/18.62 \angle 57.5°$
$= 22.29 \angle -57.5°$ A.
The total power dissipated $= 3 \times 22.29^2 \times 10 = 14\,905$ W.   (*Ans.*)
The ratio (delta power)/(star power) $= 14\,905/4984.5 = 3.$ (*Ans.*)
(This result shows that a delta-connected balanced load takes three times as much power from a power supply as does the same load when connected in star.)

Example 2.4

A balanced load consisting of three 150 Ω resistors takes a total power of 3 kW when it is connected to a 415 V, 50 Hz three-phase supply. Calculate the line current when the load is connected in (*a*) star and (*b*) delta. Determine the ratio of the line currents.

Solution
$P_{ph} = 3000/3 = 1000$ W.
$I_{ph} = \sqrt{(1000/150)} = 2.582$ A.
(*a*) $I_L = I_{ph} = 2.582$ A.   (*Ans.*)
(*b*) $I_L = \sqrt{3} \times 2.582 = 4.472$ A.   (*Ans.*)
The ratio (delta current)/(line current) $= \sqrt{3}$.   (*Ans.*)

Relative merits of the star and delta connections

The windings of three-phase generators, or alternators, are rarely connected in delta because (*a*) the star connection provides a neutral point to which the neutral wire can be connected and (*b*) for the same generated phase voltage the star connection gives the higher line voltage.

Three-phase loads may be either star- or delta-connected. The star connection allows a neutral conductor to be used which is advantageous for unbalanced loads. The delta-connected load will dissipate three times as much power as a star-connected load for the same line current. Alternatively, the delta-connected load will dissipate the same power for a smaller line current; this reduces line losses even though it is at the expense of a higher line voltage. The reduction in the line losses means that smaller diameter conductors may be employed but the higher voltage demands better cable insulation.

The electricity supply in the UK is distributed from the generating power stations to the user via the *national grid*. The supply is generated by star-connected generators at either 22 or 33 kV and this voltage is then immediately applied to a three-phase transformer. The transformer increases the voltage to either 132, 275 or 400 kV for transmission to the locality of its destination. During its transmission the voltage may be changed to/from any of these three values. At its destination the transmitted voltage is applied to another three-phase transformer and here it is transformed down to either 33 or 11 kV. The former voltage is used for the direct supply of certain very large electricity consumers, and the latter voltage is used to distribute the supply around a local area. Some consumers take their supply at 11 kV but for most, including (one hopes!) the domestic user, the supply is first further transformed down to 415/240 V. The use of such large voltages in the electricity supply distribution network minimizes power losses and so reduces costs.

Unbalanced loads

Single-phase loads, such as the electricity supply to the domestic consumer, are connected between a line conductor and the neutral conductor and the supply is at 240 V. The total load forecast is divided into, as near as possible, three equal parts, and one part is connected to each phase of the supply. The domestic load is always at, or very near to, unity power factor. In general, the three single-phase loads will not be equal to one another, since the demand fluctuates continuously, and so the three line currents are not of equal amplitude. The sum of the three line currents is therefore not equal to zero and so a fourth, neutral, conductor must be provided. The current I_N that flows in the neutral conductor is equal to the sum of the line currents, i.e.

$$I_N = I_R + I_Y + I_B \qquad (2.9)$$

Power

The total power dissipated by an unbalanced load is equal to the sum of the powers dissipated in each phase but it is not equal to $\sqrt{(3)}V_L I_L$ which only applies to a balanced load. For each phase,

$$\text{power} = V_{ph} I_{ph} \cos \phi \quad \text{W} \qquad (2.10)$$

The overall power factor of an unbalanced load is given by equation (2.11), i.e.

$$\text{Overall power factor} = (\text{total kW})/(\text{total kVA}) \qquad (2.11)$$

Example 2.5

A 415 V, 50 Hz three-phase supply is applied to the delta-connected load shown in Fig. 2.18. Calculate (*a*) the total power dissipated and the overall power factor and (*b*) the total power dissipated and the overall power factor if the load were star-connected.

Solution
(*a*) For the delta connection $V_{ph} = V_L = 415$ V.
$I_{RY} = 415/[\sqrt{(100^2 + 100^2)}] \angle \tan^{-1}(100/100) = 415/141.4 \angle 45°$ A.
$P_R = (415/141.4)^2 \times 100 = 861.39$ W.

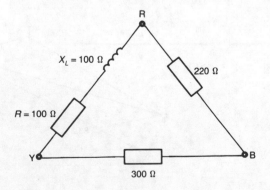

Fig. 2.18

$I_{YB} = 415/300$ A. $P_Y = (415/300)^2 \times 300 = 574.08$ W.
$I_{BR} = 415/220$ A. $P_B = (415/220)^2 \times 220 = 782.84$ W.
Total power dissipated $= 861.39 + 574.08 + 782.84 = 2218.3$ W.   (*Ans.*)
Total kVA $= 415 \times [(415/141.4) + (415/300) + (415/220)] = 2574.9$ kVA.
Therefore, overall power factor $= 2218.3/2574.9 = 0.862$.   (*Ans.*)
(*b*) $V_{ph} = 415/\sqrt{3} = 240$ V. $|I_R| = 240/141.4$ A
$P_R = (240/141.4)^2 \times 100 = 288.09$ W.
$I_Y = 240/300$ A. $P_Y = (240/300)^2 \times 300 = 192$ W.
$I_B = 240/220$ A. $P_B = (240/220)^2 \times 220 = 261.82$ W.
Total power $= 288.09 + 192 + 261.82 = 741.91$ W.   (*Ans.*)
Total kVA $= 240[(240/141.1) + (240/300) + (240/220)] = 861.17$ kVA.
Therefore, overall power factor $= 741.91/861.17 = 0.862$.   (*Ans.*)

Example 2.6

A three-phase 50 Hz four-wire three-phase system supplies the following loads:
(i) single-phase loads of 12, 18 and 20 kW, and (ii) a star-connected three-phase load of 20 kW at 0.75 lagging power factor. Calculate (*a*) the three line currents and (*b*) the current in the neutral conductor.

Solution

The three single-phase loads are $I'_R = 12\,000/240 = 50$ A,
$\quad I'_Y = 18\,000/240 = 75$ A and $I'_B = 20\,000/240 = 83.33$ A.
The three-phase load takes 20 kW power and hence
$\quad 20\,000 = 3 \times 240 \times 0.75 I_{ph}$.
$\quad I_{ph} = 20\,000/(3 \times 240 \times 0.75) = 37$ A. $\phi = \cos^{-1}0.75 = 41.4°$.
The phasor diagram is shown in Fig. 2.19. Resolving I_R into its in-phase and quadrature components with respect to V_R gives

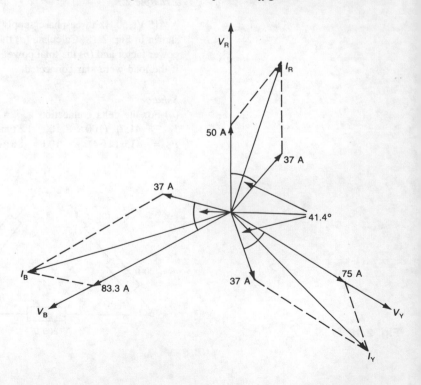

Fig. 2.19

37 cos 41.4° = 27.75 A, and 37 sin 41.4° = 24.47 A. Therefore

$$I_R = \sqrt{[(50 + 27.75)^2 + 24.47^2]} \angle \tan^{-1}(-24.47/77.75)$$
$$= 81.5 \angle -17.5° \text{ A.} \quad (Ans.)$$

Similarly, $I_Y = \sqrt{[(75 + 27.75)^2 + 24.47^2]} \angle \tan^{-1}(-24.47/102.75)$
$$= 105.62 \angle -13.4° \text{ A, relative to } V_Y,$$
$$= 105.62 \angle -133.4° \text{ A, relative to } V_R. \quad (Ans.)$$

Also $I_B = \sqrt{[(83.33 + 27.75)^2 + 24.47^2]} \angle \tan^{-1}(-24.47/111.08)$
$$= 113.74 \angle \tan^{-1} -12.4° \text{ A, relative to } V_B,$$
$$= 113.74 \angle -252.4° \text{ A, relative to } V_R. \quad (Ans.)$$

Resolving each line current into its horizontal and vertical components gives

Horizontal: 81.51 cos (−17.5°) + 105.62 cos (−133.4°)
$$+ \ 113.74 \cos (-252.4°) = 77.74 - 72.57 - 34.39$$
$$= -29.22 \text{ A.}$$

Vertical: 81.5 sin (−17.5°) + 105.62 sin (−133.4°)
$$+ \ 113.74 \sin (-252.4°) = -24.51 - 76.74 + 108.42$$
$$= 7.17 \text{ A.}$$

The neutral current is $I_N = \sqrt{(29.22^2 + 7.17^2)} \angle \tan^{-1}(7.17/-29.22)$
$$= 30.09 \angle 166.2° \text{ A.} \quad (Ans.) \text{ (relative to } V_R)$$

Measurement of three-phase power

The power dissipated in a three-phase system may be measured using one, two or three wattmeters.

The one-wattmeter method

A single wattmeter can be employed to measure the total power dissipated in a *balanced* three-phase load. The necessary connections when measuring the power in a star-connected load are shown in Fig. 2.20. The current coil of the wattmeter is connected in series with one line (the red line in the figure) and its voltage coil is connected between that line and the neutral point of the load. Since the load is balanced the total power dissipated is equal to three times the power P_A indicated by the wattmeter. Therefore

$$\text{total power } P_T = 3P_A \tag{2.12}$$

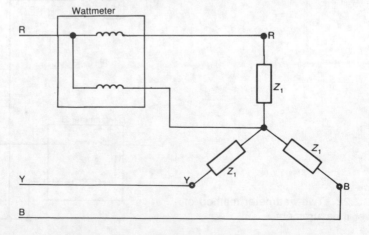

Fig. 2.20 One-wattmeter method of power measurement

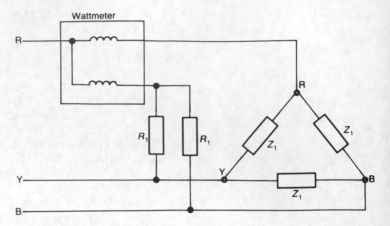

Fig. 2.21 Simulated neutral point

A disadvantage of the one-wattmeter method of power measurement is that it requires access to the neutral point of the load. Should this point not be accessible or if the load is delta-connected a virtual neutral point will need to be created. One method of doing this is shown by Fig. 2.21.

The two-wattmeter method

The total power dissipated in a three-phase load, balanced or unbalanced, can be measured using two wattmeters connected as shown by Fig. 2.22. Each of the wattmeters has its current coil connected in series with one of the line conductors and has its voltage coil connected between that line and the third line conductor. Although a star-connected load has been shown the method works equally well with a delta-connected load.

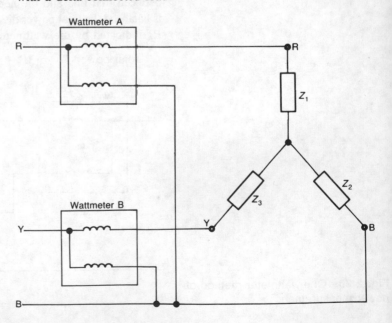

Fig. 2.22 Two-wattmeter method of power measurement

The voltage across the voltage coil of wattmeter A is V_{RB} and the voltage applied to the voltage coil of wattmeter B is V_{YB}. If the instantaneous powers in the two wattmeters are p_A and p_B then

$$p_A + p_B = v_{RB}i_R + v_{YB}i_Y$$
$$= (v_R - v_B)i_R + (v_Y - v_B)i_Y$$
$$= v_R i_R + v_Y i_Y - v_B(i_R + i_Y)$$

But for a three-wire system $i_B = i_R + i_Y$ and hence

$$p_A + p_B = v_R i_R + v_Y i_Y + v_B i_B$$

or

$$p_A + p_B = p_R + p_Y + p_B \qquad (2.13)$$

Thus, the sum of the instantaneous powers in the two wattmeters is equal to the sum of the instantaneous powers dissipated in the three phases of the load. The wattmeters are unable to follow the instantaneous values of the powers and instead indicate the *mean* powers. Hence

$$P_A + P_B = P_R + P_Y + P_B \qquad (2.14)$$

This result is valid for all loads, balanced or unbalanced, that are supplied by a three-wire, three-phase, system.

The indications of the two wattmeters will only be the same if the load is balanced and of unity power factor. If overall load power factor is less than 0.5 one of the wattmeters will indicate a negative value of power. The total power dissipated in the load is then equal to the difference between the two indicated values.

Measurement of load power factor

The two-wattmeter method of measuring power can also be used to find the power factor of a balanced load. Reference to the phasor diagram given in Fig. 2.23 shows that the phase angle between V_{RB} and I_R is $30° - \phi$, where $\cos \phi$ is the power factor. It can also be seen that the phase angle between V_{YB} and I_Y is $30° + \phi$.

Wattmeter A indicates a power P_A of $V_{RB}I_R \cos (30° - \phi)$.

Wattmeter B indicates a power P_B of $V_{YB}I_Y \cos (30° + \phi)$.

Therefore

$$P_A/P_B = [\cos(30° - \phi)]/[\cos(30° + \phi)]$$
$$= (\cos 30° \cos \phi + \sin 30° \sin \phi)/(\cos 30° \cos \phi - \sin 30° \sin \phi)$$

Dividing top and bottom of the right-hand side of the equation by $\cos 30° \cos \phi$ gives

$$P_A/P_B = (1 + \tan 30° \tan \phi)/(1 - \tan 30° \tan \phi)$$
$$= (1 + \tan \phi/\sqrt{3})/(1 - \tan \phi/\sqrt{3})$$

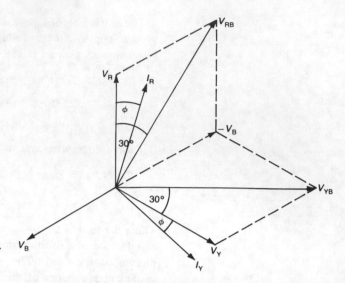

Fig. 2.23 Calculation of load power factor

Hence

$$P_A(1 - \tan \phi / \sqrt{3}) = P_B(1 + \tan \phi / \sqrt{3})$$

$$P_A - P_B = (P_A + P_B) \tan \phi / \sqrt{3}$$

Therefore

$$\tan \phi = \sqrt{3}[(P_A - P_B)/(P_A + P_B)] \qquad (2.15)$$

Example 2.7

The two-wattmeter method of three-phase power measurement is used to determine the power factor of a balanced load. The indications of the two wattmeters are 5 and 2 kW. (*a*) Calculate the power factor of the load and the total power dissipated. (*b*) Repeat the calculation if wattmeter B indicates −2 kW.

Solution
(*a*) Tan $\phi = \sqrt{3}(3/7) = 0.742$, and hence $\phi = 36.6°$.
The power factor = cos 36.6° = 0.8.   (*Ans.*)
The total power dissipated = 7 kW.   (*Ans.*)
(*b*) Tan $\phi = \sqrt{3}(7/3) = 4.04$ and $\phi = 76.1°$.
The power factor = cos 76.1° = 0.24.   (*Ans.*)
The total power dissipated = 3 kW.   (*Ans.*)

The three-wattmeter method

The total power dissipated in a load, balanced or unbalanced, can be measured using the three-wattmeter method shown in Fig. 2.24. The current coil of one wattmeter is connected in series with each line and the voltage coils are connected between that line and the neutral

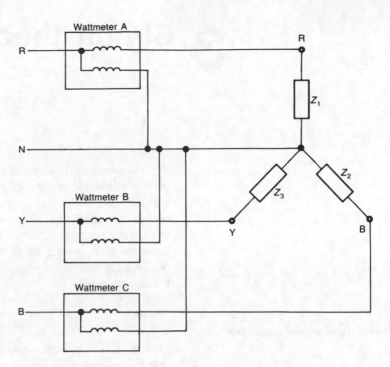

Fig. 2.24 Three-wattmeter method of power measurement

conductor. The total power dissipated is then the sum of the powers indicated by the three wattmeters, i.e.

$$P_T = P_R + P_Y + P_B \tag{2.16}$$

The three-wattmeter method of power measurement must be used to determine the total power dissipated in a four-wire star-connected load.

3 Circuit theorems

The calculation of the currents in, and the voltages across, the various parts of an electric circuit can always be carried out using Kirchhoff's current and voltage laws. Very often, however, a considerable reduction in the amount of work involved can be achieved if one, or more, of a number of circuit theorems is/are employed. The circuit theorems to be considered in this chapter are Thevenin's theorem, Norton's theorem, the superposition theorem and the maximum power transfer theorem.

Current and voltage sources

Most sources of electrical energy have an internal resistance and are able to provide both a current and a voltage to an external load. The current and voltage may be either a.c. or d.c. depending upon the source, but in either case the values of the supplied current and the terminal voltage may vary if the impedance of the load is changed. Some sources produce an output current and/or voltage that is determined solely by the source itself, the prime examples of this being a primary or secondary cell battery and the mains electricity supply. Other sources, known as dependent sources, provide an output current and/or voltage that is determined by another applied current or voltage. A bipolar transistor is an example of a dependent current source since its collector current is set by its base/emitter voltage, and an op-amp is an example of a dependent voltage source since its output voltage is determined by the voltage applied between its inverting and non-inverting input terminals.

Any practical source may be considered to be either a voltage or a current generator. A voltage source consists of a constant voltage generator in series with its internal resistance, as shown by Fig. 3.1(*a*), while a current source consists of a constant current generator connected in parallel with the internal resistance (Fig. 3.1(*b*)). The ideal voltage source provides a terminal voltage that remains at a constant value as the current taken by the load varies. This means, of course, that the ideal voltage source must have zero internal resistance so that there is no internal voltage drop, and then a constant terminal voltage will be maintained. The ideal current source will have an infinite internal resistance so that the current that flows from its output terminals remains constant no matter how much the impedance of the load may change. In practice, the requirements for a constant

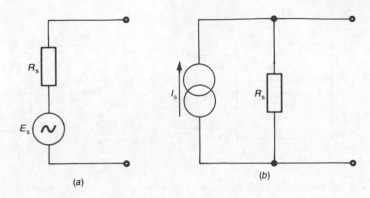

Fig. 3.1 (a) Voltage source; (b) current source

current, or a constant voltage, generator, cannot be satisfied. However, the principle shows that a voltage source should have a low internal resistance, and a current source should have a high internal resistance, relative to the value of the external load impedance.

Example 3.1

(a) A voltage source has an e.m.f. of 6 V and an internal resistance of (i) $5\,\Omega$, and (ii) $1000\,\Omega$. Calculate the change in the terminal voltage in each case if the source is connected to a load whose resistance can be varied from 100 to $1000\,\Omega$.

(b) A constant current source of 5 mA has an internal resistance of (i) $1000\,\Omega$, and (ii) $20\,000\,\Omega$. Calculate the change in the load current as the load resistance varies from 100 to $1000\,\Omega$.

Solution

(a)(i) The terminal voltage varies from $(6 \times 100)/(5 + 100) = 5.714\,\text{V}$ to $(6 \times 1000)/(5 + 1000) = 5.97\,\text{V}$. *(Ans.)*

(ii) The terminal voltage varies from $(6 \times 100)/(1000 + 100) = 0.546\,\text{V}$ to $(6 \times 1000)/(1000 + 1000) = 3\,\text{V}$. *(Ans.)*

(b)(i) The load current varies from $(5 \times 1000)/(1000 + 100) = 4.55\,\text{mA}$ to $(5 \times 1000)/(1000 + 1000) = 2.5\,\text{mA}$. *(Ans.)*

(ii) The load current varies from $(5 \times 20\,000)/(100 + 20\,000) = 4.975\,\text{mA}$ to $(5 \times 20\,000)/(1000 + 20\,000) = 4.762\,\text{mA}$. *(Ans.)*

Thevenin's theorem

Thevenin's theorem states that

the current flowing in an impedance Z_L connected across the output terminals of a linear network† will be the same as the current that would flow if the load impedance Z_L were connected to a voltage source of e.m.f. V_{oc} and internal impedance Z_{oc}. V_{oc} is the voltage that appears across the output terminals when they are open-circuited and Z_{oc} is the output impedance of the network when all of its internal voltage and/or current sources have been replaced by their internal impedances.

---

† A linear network is one in which all impedances obey Ohm's law.

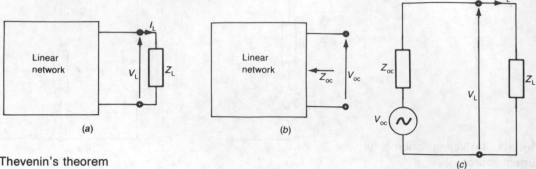

Fig. 3.2 Thevenin's theorem

This statement of Thevenin's theorem can be summarized by Fig. 3.2. Any point in the network may be considered to be the output terminals for the application of Thevenin's theorem and the theorem may be applied to different parts of the network as many times as required.

Example 3.2

For the network shown in Fig. 3.3 calculate the current that flows in the 1 kΩ load resistor R_L using (*a*) Ohm's law, (*b*) Kirchhoff's laws and (*c*) Thevenin's theorem.

Solution

(*a*) The total resistance connected across the terminals of the d.c. voltage source is

$$R = 1000 + (2200 \times 4300)/(2200 + 4300) = 2455.4 \,\Omega.$$

The current I taken from the source is $I = 6/2455.4$ A, and hence the current flowing in the load resistor is

$$I_L = (6/2455.4) \times 2200/(2200 + 4300) = 827 \,\mu\text{A}. \quad (Ans.)$$

(*b*) From Fig. 3.4(*a*)

$$6 = (1000 + 2200)I_1 - 2200I_2 = 3200I_1 - 2200I_2 \quad (3.1)$$
$$0 = -2200I_1 + (2200 + 3300 + 1000)I_2 = -2200I_1 + 6500I_2 \quad (3.2)$$

From equation (3.2), $I_1 = 6500I_2/2200$ and substituting this value into equation (3.1) gives

$$6 = 3200(6500/2200)I_2 - 2200I_2, \text{ or } I_2 = 6/7254.6$$
$$= 827 \,\mu\text{A}. \quad (Ans.)$$

(*c*) With the load resistance disconnected from the network the voltage V_{oc} that appears across the output terminals is

$$V_{oc} = (6 \times 2200)/(1000 + 2200) = 4.125 \text{ V}.$$

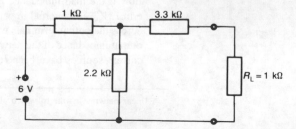

Fig. 3.3

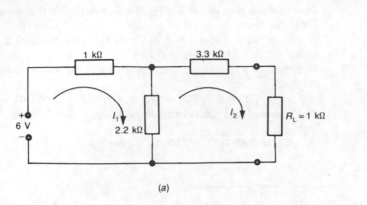

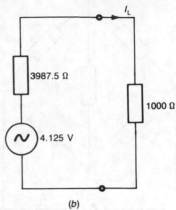

(a)

(b)

Fig. 3.4

The output resistance R_{oc} of the network, with the 6 V d.c. source replaced by its $0\,\Omega$ internal resistance, is

$R_{oc} = 3300 + (1000 \times 2200)/(1000 + 2200) = 3987.5\,\Omega$.

The Thevenin equivalent circuit of the network is given in Fig. 3.4(b). From this circuit,

$I_L = 4.125/(3987.5 + 1000) = 827\,\mu A$. (Ans.)

It may seem from this example that very little advantage, if any, is gained by the use of Thevenin's theorem to solve a circuit problem but when it is applied to more complex networks its advantages soon become apparent.

Example 3.3

Use Thevenin's theorem to calculate the voltage across the capacitor in the circuit given in Fig. 3.5.

Solution

With the capacitor disconnected from the circuit the voltage V_{oc} that appears across the open-circuited output terminals is

$V_{oc} = (1 \times 2000)/2600 = 0.77\,V$.

With the a.c. voltage source replaced by its internal impedance of $0\,\Omega$ the output impedance Z_{oc} of the circuit is

$Z_{oc} = (2000 \times 600)/2600 = 461.54\,\Omega$.

The Thevenin equivalent circuit is shown by Fig. 3.6. The impedance of this circuit is

$Z = \sqrt{(461.54^2 + 1000^2)} \angle \tan^{-1}(-1000/461.54)$
$= 1101.4 \angle -65.2°\,\Omega$.

The current flowing in the circuit is

$I = 0.77\angle 0°/1101.4\angle -65.2° = 6.99 \times 10^{-4} \angle 65.2°\,A$.

The voltage across the capacitor $= 6.99 \times 10^{-4} \angle 65.2° \times 1000 \angle -90°$
$= 0.699 \angle -24.8°\,V$. (Ans.)

Fig. 3.5

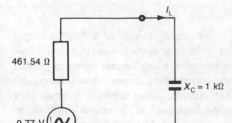

Fig. 3.6

Example 3.4

Use Thevenin's theorem to calculate the current flowing in the 200 Ω resistor in the bridge network given in Fig. 3.7.

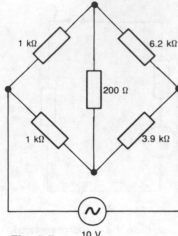

Fig. 3.7

Solution

The circuit has been redrawn in Fig. 3.8(*a*) with the 200 Ω resistor removed. From this figure

$V_A = (10 \times 3.9)/4.9 = 7.96$ V, and $V_B = (10 \times 6.2)/7.2 = 8.61$ V.

Hence $V_{oc} = 8.61 - 7.96 = 0.65$ V.

With the 10 V a.c. voltage source replaced by its 0 Ω internal resistance (Fig. 3.8(*b*)),

$R_{oc} = [(1 \times 6.2)/7.2] + [(1 \times 3.9)/4.9] = 1.657$ kΩ.

The Thevenin equivalent circuit of the bridge is shown in Fig. 3.8(*c*) and from this

$I_{200} = 0.65/1857 = 350\,\mu$A.   (*Ans.*)

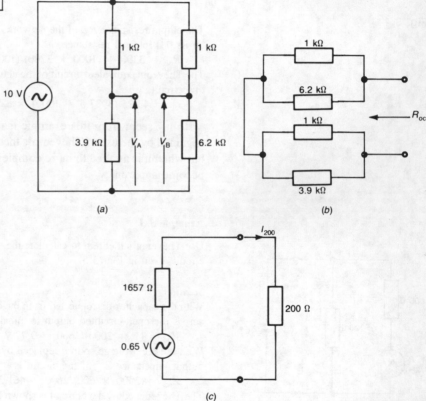

(a)   (b)

(c)

Fig. 3.8

Norton's theorem

Norton's theorem states that

> the current flowing in an impedance Z_L connected across the output terminals of a linear network is the same as the current that would flow if the impedance Z_L were connected across a current source in parallel with an impedance. The output current of the current source is the current I_{sc} that would flow in the short-circuited output terminals of the network and the impedance is the output impedance of the network with all the internal current and voltage sources replaced by their internal impedances.

Norton's theorem is summarized by the circuits given in Fig. 3.9.

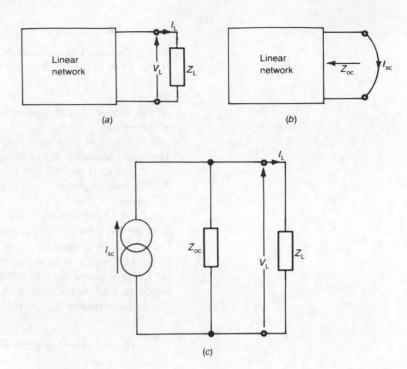

Fig. 3.9 Norton's theorem

Example 3.5

Calculate, using Norton's theorem, the current flowing in the load resistor R_L in the circuit shown in Fig. 3.3.

Solution

With the output terminals of the network short-circuited the total resistance R connected across the voltage source is

$$R = 1000 + (2200 \times 3300)/(2200 + 3300) = 2320\,\Omega.$$

The current I taken from the voltage source is $I = 6/2320$ A, and the current I_{sc} flowing in the short-circuit is

$$I_{sc} = (6/2320) \times 2200/(2200 + 3300) = 1.0345\,\text{mA}.$$

As before, the output resistance R_{oc} of the network is $R_{oc} = 3987.5\,\Omega$ and hence the Norton equivalent circuit is as shown by Fig. 3.10. From this circuit,

$$I_L = 1.0345 \times 3987.5/4987.5\,\text{mA} = 827\,\mu\text{A}. \quad (Ans.)$$

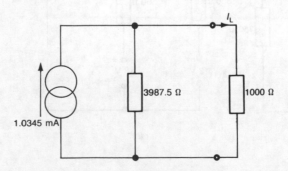

Fig. 3.10

Example 3.6

Use (*a*) Thevenin's theorem and (*b*) Norton's theorem to calculate the current I_L that flows in the 750 Ω load resistor in the circuit shown in Fig. 3.11.

Solution

(*a*) The voltage V_{oc} that appears across the open-circuited output terminals of the network is

$$V_{oc} = 18 - 560[(18 - 12)/860] = 14.093 \text{ V}.$$

The output resistance R_{oc} of the network is

$$R_{oc} = (300 \times 560)/860 = 195.4 \text{ Ω}.$$

The Thevenin equivalent circuit of the network is given by Fig. 3.12(*a*). From this circuit,

$$I_L = 14.093/945.4 = 14.9 \text{ mA}. \quad (Ans.)$$

(*b*) If the 750 Ω load resistor is replaced by a short-circuit the circuit shown in Fig. 3.12(*b*) is obtained. From this circuit the current supplied by:

(i) the 12 V source is $I_1 = 12/300 = 40$ mA.

(ii) the 18 V source is $I_2 = 18/560 = 32.14$ mA.

Hence the total short-circuit current $I_{sc} = I_1 + I_2 = 72.14$ mA.

The output resistance $R_{oc} = 195.4$ Ω.

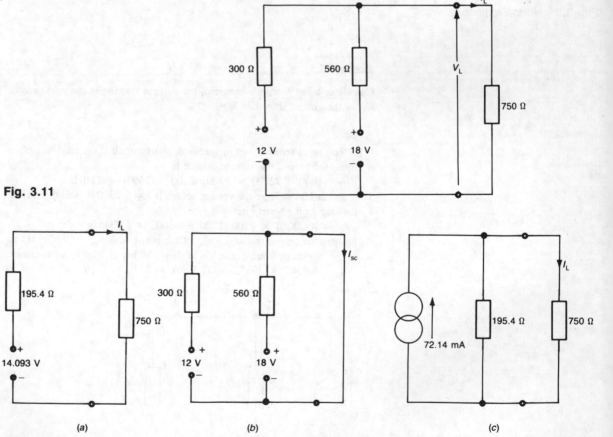

Fig. 3.11

(a)   (b)   (c)

Fig. 3.12

The Norton equivalent circuit is given by Fig. 3.12(c). From this circuit,
$$I_L = 72.14 \times [195.4/(195.4 + 750)] = 14.9 \text{ mA}. \quad (Ans.)$$

Converting from Thevenin to Norton

Sometimes the reduction of a complex network results in a combination of both the Thevenin and the Norton equivalent circuits and then it is necessary to convert one or other of the circuits to obtain either just a Thevenin equivalent circuit or just a Norton equivalent circuit. Figure 3.13(a) shows the generalized Thevenin equivalent circuit. If Norton's theorem is applied to this circuit $I_{sc} = V_{oc}/Z_{oc}$ and $Z_{oc} = Z_{oc}$, giving the Norton equivalent circuit shown in Fig. 3.13(b). If, Thevenin's theorem is now applied to this circuit, $V_{oc} = (V_{oc}/Z_{oc}) \times Z_{oc} = V_{oc}$, and $Z_{oc} = Z_{oc}$ which gives Fig. 3.13(a) again.

Example 3.7

Determine the Thevenin equivalent of the circuit shown in Fig. 3.14 and hence calculate the power dissipated in the 1000 Ω load resistor. Assume that the two generators operate at the same frequency and are in phase with one another.

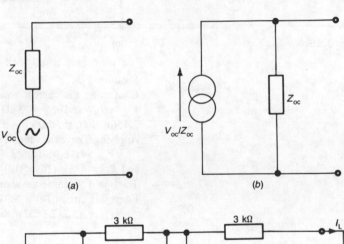

Fig. 3.13 Equivalent circuits: (a) Thevenin's theorem; (b) Norton's theorem

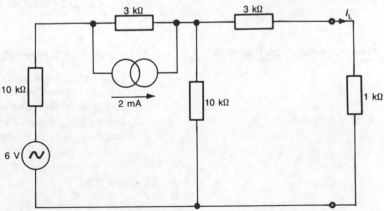

Fig. 3.14

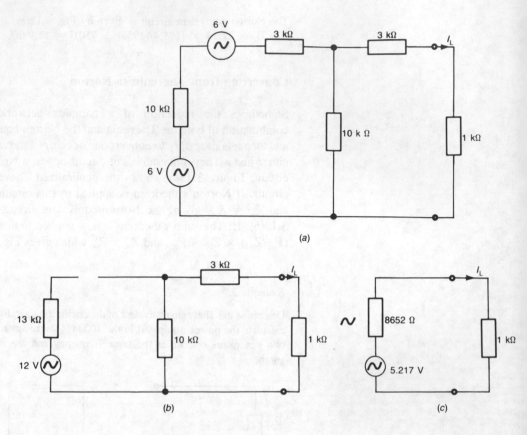

Fig. 3.15

(a)

(b)

(c)

Solution

Converting the current source into its equivalent Thevenin circuit gives $V_{oc} = 6\,V$ and $R_{oc} = 3\,k\Omega$. Hence the circuit can be redrawn as in Fig. 3.15(*a*). In turn, this circuit can be further simplified to give Fig. 3.15(*b*). Applying Thevenin's theorem to the output terminals of this network gives

$$V_{oc} = (12 \times 10)/23 = 5.217\,V,$$

and $R_{oc} = 3 + (10 \times 13)/(10 + 13)\,k\Omega = 8.652\,k\Omega$.

The final Thevenin equivalent circuit for the network is given in Fig. 3.15(*c*). From this circuit $I_L = 5.317/9652\,A$ and the load power is

$$P_L = (5.217/9652)^2 \times 1000 = 292.2\,\mu W. \quad (Ans.)$$

The superposition theorem

The *superposition theorem* states that

> the current at any point in a *linear* network that contains two, or more, voltage or current sources, is equal to the sum of the currents that would flow at that point if only one source were considered at a time, all other sources being replaced by an impedance whose value is equal to the internal impedance of the source.

Example 3.8

Calculate the current flowing in the 20 Ω resistor in the circuit of Fig. 3.16 if the two voltage sources operate at the same frequency and their e.m.f.s

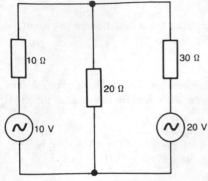

Fig. 3.16

are (*a*) in phase, and (*b*) in anti-phase, with one another. Check the answer using Norton's theorem.

Solution
(i) Replace the left-hand voltage source with a short-circuit to give the circuit shown in Fig. 3.17(*a*). From this circuit the total resistance R_1 connected across the 20 V source is
$$R_1 = 30 + (20 \times 10)/30 = 36.67 \,\Omega.$$
Therefore, $I_1 = (20/36.67) \times (10/30) = 181.8 \,\text{mA}.$
(ii) Replace the right-hand voltage source by a short-circuit to give the circuit shown by Fig. 3.17(*b*). The total resistance R_2 connected across the 10 V voltage source is
$$R_2 = 10 + (20 \times 30)/50 = 22 \,\Omega.$$
Therefore $I_2 = (10/22) \times (30/50) = 272.73 \,\text{mA}.$
(*a*) When the two voltage sources are in phase with one another
$$I_{20} = I_1 + I_2 = 181.8 + 272.73 = 454.5 \,\text{mA}. \quad (Ans.)$$
(*b*) When the sources are in anti-phase with one another
$$I_{20} = I_2 - I_1 = 90.93 \,\text{mA}. \quad (Ans.)$$
Using Norton's theorem the first step is to short-circuit the output terminals of the network to obtain the circuit shown in Fig. 3.17(*c*). From this circuit the current supplied by the 10 V source is $10/10 = 1$ A. The current supplied by the 20 V source is $20/30 = 0.67$ A.
(*a*) With the voltage sources in phase the current I_{sc} in the short-circuit is 1.67 A. The output resistance R_{oc} of the network is
$$R_{oc} = (30 \times 10)/40 = 7.5 \,\Omega.$$
The Norton equivalent circuit is shown in Fig. 3.17(*d*). From this circuit,
$$I_{20} = 1.67 \times (7.5/27.5) = 455.5 \,\text{mA}. \quad (Ans.)$$
(*b*) With the voltage sources in anti-phase the short-circuit current $I_{sc} = 0.33$ A. Hence
$$I_{20} = 0.33 \times (7.5/27.5) = 90 \,\text{mA}. \quad (Ans.)$$

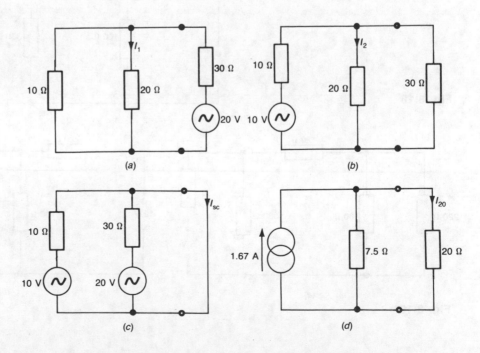

Fig. 3.17

Example 3.9

Use the superposition theorem to calculate the current flowing in each of the resistors in the circuit given by Fig. 3.18.

Solution

(*a*) Short-circuit the left-hand voltage source. The resulting circuit is then shown by Fig. 3.19(*a*). The resistance 'seen' looking to the left of the 330 Ω resistor is

$$120 + (220 \times 100)/320 = 188.75 \, \Omega.$$

Hence the total resistance across the terminals of the 6 V source is

$$56 + (330 \times 188.75)/(330 + 188.75) = 176 \, \Omega.$$

Therefore $I'_{56} = 6/176 = 34.1$ mA.

$I'_{330} = 34.1 \times 188.75/(188.75 + 330) = 12.41$ mA.

$I'_{120} = 34.1 - 12.41 = 21.69$ mA.

$I'_{100} = 21.69 \times 220/(220 + 100) = 14.91$ mA.

$I'_{220} = 21.69 - 14.91 = 6.78$ mA.

(*b*) With the right-hand voltage source short-circuited the circuit is as shown by Fig. 3.19(*b*). The resistance 'seen' looking to the right of the 100 Ω resistor is

$$120 + (330 \times 56)/(330 + 56) = 167.88 \, \Omega.$$

Hence the total resistance across the terminals of the 12 V source is

$$220 + (100 \times 167.88)/(267.88) = 282.67 \, \Omega.$$

Therefore, $I''_{220} = 12/282.67 = 42.45$ mA.

$I''_{100} = 42.45 \times 167.88/(167.88 + 100) = 26.6$ mA.

$I''_{120} = 42.45 - 26.6 = 15.85$ mA.

$I''_{330} = 15.85 \times 56/(330 + 56) = 2.3$ mA.

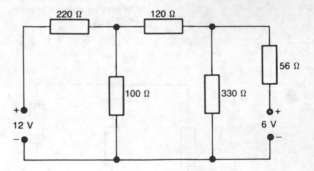

Fig. 3.18

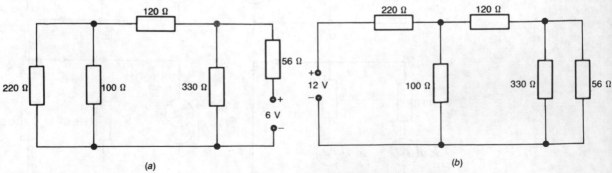

(a)   (b)

Fig. 3.19

$I_{56}'' = 15.85 - 2.3 = 13.55 \text{ mA}$. Therefore
$I_{220} = 42.45 - 6.78 = 35.67 \text{ mA}$. (*Ans.*)
$I_{100} = 14.91 + 26.6 = 41.51 \text{ mA}$. (*Ans.*)
$I_{120} = 21.69 - 15.85 = 5.84 \text{ mA}$. (*Ans.*)
$I_{330} = 12.41 + 2.3 = 14.71 \text{ mA}$. (*Ans.*)
$I_{56} = 34.1 - 13.55 = 20.55 \text{ mA}$. (*Ans.*)

Maximum power transfer theorem

Many instances occur in electronic and radio engineering where a voltage source is connected to a load and the maximum possible power is required to be transferred from the source to the load. For this to happen there must be a particular relationship between the source and load impedances. Figure 3.20 shows a voltage source of e.m.f. 10 V and internal resistance 10 Ω connected to a variable resistor R. The current I flowing in the circuit is $I = 10/(10 + R)$ amperes, and the power dissipated in the load resistor is $P_L = [10/(10 + R)]^2 R$ watts.

If the resistance of the load is gradually increased, starting from 0 Ω, it will be found that the load power increases at first, then reaches a maximum value, and thereafter it decreases. For the maximum power to be transferred from the source to the load the load resistance should be set to the value at which it dissipates the maximum possible power. Suppose that the load resistance R is increased in 2 Ω steps from 0 to 20 Ω. The values then of both the current flowing, and the power dissipated, in the load are given in Table 3.1 and are shown plotted in Fig. 3.20(*b*).

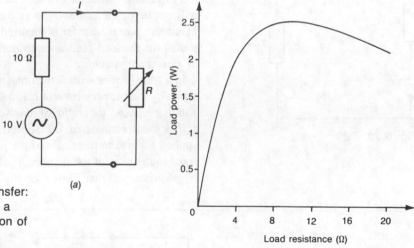

Fig. 3.20 Maximum power transfer: (*a*) voltage source connected to a variable load resistor; (*b*) variation of load power

Table 3.1

R	(Ω)	0	2	4	6	8	10	12	14	16	18	20
I	(A)	1	0.833	0.714	0.625	0.556	0.5	0.455	0.417	0.385	0.357	0.333
P	(W)	0	1.389	2.041	2.344	2.469	2.5	2.479	2.431	2.367	2.296	2.222

It can be seen that the maximum power is dissipated in the load when its resistance has the same value as the source resistance. This leads to one statement of the maximum power transfer theorem, i.e. 'for the maximum power to be transferred from a purely resistive source to a purely resistive load, the resistance R_L of the load must be equal to the resistance R_s of the source'.

$$R_s = R_L \tag{3.3}$$

Example 3.10

For the circuit shown in Fig. 3.21 determine (a) the value of the load resistance R_L that will dissipate the maximum power, and (b) the value of this maximum power.

Solution

Applying Thevenin's theorem to the output terminals of the circuit gives
$$V_{oc} = (12 \times 20)/32 = 7.5\,V,$$
and $R_{oc} = (12 \times 20)/32 = 7.5\,k\Omega.$
The Thevenin equivalent circuit is shown in Fig. 3.22 and from this:
(a) $R_L = 7.5\,k\Omega.$ (*Ans.*)
and (b) $P_{(max)} = 3.75^2/(7.5 \times 10^3) = 1.875\,mW.$ (*Ans.*)

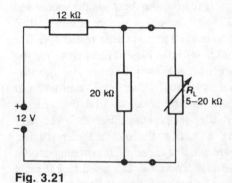

Fig. 3.21

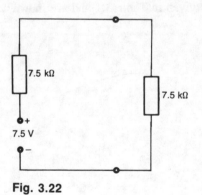

Fig. 3.22

Use of a transformer for resistance matching

Very often the resistance of a load that is connected to a voltage source will not be of the same value as the source resistance and then, if maximum power transfer is required, a transformer must be used to transform the load resistance to the required value. This is known as *resistance matching*.

In a transformer with a ferromagnetic core almost all of the flux set up by the primary current I_p links with the turns of the secondary winding. Then, the ratio of the secondary voltage V_s appearing across the terminals of the secondary winding to the voltage V_p applied across the terminals of the primary winding, is very nearly equal to the ratio of the number N_s of secondary turns to the number N_p of primary turns. Thus (see Fig. 3.23),

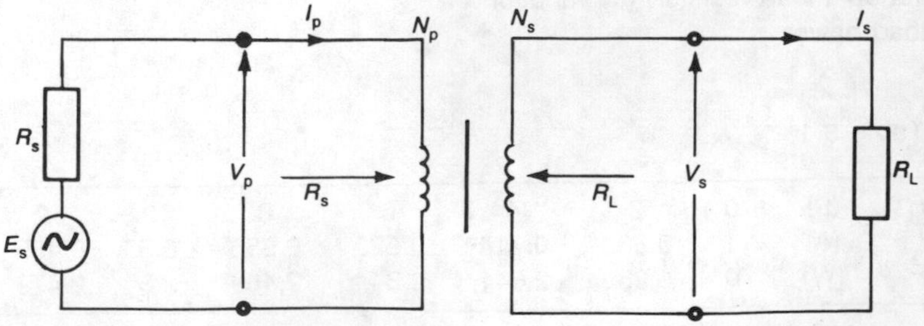

Fig. 3.23 Resistance matching

$$V_s/V_p = N_s/N_p \tag{3.4}$$

If the power losses in the transformer are negligibly small so that the output power is equal to the input power then $I_p V_p = I_s V_s$ and

$$I_p/I_s = V_s/V_p = N_s/N_p \tag{3.5}$$

This means that the current ratio of a transformer is the inverse of its voltage ratio.

Now $I_p = V_p/R_s$ and $I_s = V_s/R_L$, assuming that the winding resistances are small compared with R_s and R_L. Hence substituting into equation (3.5),

$$(V_p/R_s)/(V_s/R_L) = N_s/N_p$$

$$(V_p R_L)/(V_s R_s) = N_s/N_p$$

and therefore,

$$R_L/R_s = (N_s/N_p)^2 \tag{3.6}$$

Thus the resistance ratio of a transformer is equal to the square of its turns ratio.

Example 3.11

A transformer is to be used to match a 15 V 75 Ω a.c. voltage source to a 1875 Ω load. Calculate (a) the necessary turns ratio for the transformer, (b) the load power assuming zero transformer losses and (c) the load power if the source were to be directly connected to the load.

Solution
(a) $N_s/N_p = \sqrt{(1875/75)} = 5:1$.   (*Ans.*)
(b) $P_L = 7.5^2/75 = 0.75\,\text{W}$.   (*Ans.*)
(c) $I = 15/(75 + 1875) = 7.69\,\text{mA}$.
$P_L = (7.69 \times 10^{-3})^2 \times 1875 = 0.11\,\text{W}$.   (*Ans.*)

In practice, of course, there are always some transformer losses and so the true load power will always be less than calculated.

If the source and/or the load impedance has both resistive and reactive components then the maximum power will be transferred to the load if (i) the load resistance is of the same value as the source resistance, and (ii) the load reactance has the same magnitude as the source reactance but is of the opposite sign. If, for example, a voltage source has an output impedance that consists of a 20 Ω resistance in series with a capacitive reactance of 12 Ω maximum power will be delivered to a load of 20 Ω resistance in series with an inductive reactance of 12 Ω.

It may often be possible to connect a suitable value of reactance in series with the load in order to obtain the required condition for maximum power transfer. If not, the maximum available load power can be obtained by using a transformer to match the *magnitudes* of the source and the load impedances.

Example 3.12

A voltage source has an e.m.f. of 10 V and an internal impedance of 50 Ω resistance in series with an inductive reactance of 10 Ω. Power is to be delivered to a 25 Ω purely resistive load. Calculate (*a*) the turns ratio of the transformer required to obtain the maximum possible load power and (*b*) the value of this power.

Solution
(*a*) The magnitude of the source impedance $= \sqrt{(50^2 + 10^2)} = 51 \, \Omega$.
Turns ratio $= \sqrt{(51/25)} = 1.428 : 1$.   (*Ans.*)
(*b*) The current taken from the source is
$$10/\sqrt{[(2 \times 51)^2 + 10^2]} = 10/102.49 = 97.57 \, \text{mA}.$$
Hence, the load power $P_L = (97.57 \times 10^{-3})^2 \times 51 = 485.5 \, \text{mW}.(Ans.)$

Example 3.13

An a.c. voltage source has an e.m.f. of 2 V at an angular frequency of 10^5 rad/s and an internal impedance that consists of a 200 Ω resistor in series with a 2 mH inductor. The source is to be connected via a reactance X to the primary winding of a transformer. The secondary winding of the transformer is connected to a 5 Ω resistive load. Calculate (*a*) the value of the reactance X and the turns ratio of the transformer for the maximum load power and (*b*) the value of this power.

Solution
(*a*) The reactance X must be of the same magnitude but of the opposite polarity to the source reactance. Hence, $X_C = 200 \, \Omega = 1/10^5 C$, or
$$C = 1/(200 \times 10^5) = 50 \, \text{nF}.   (Ans.)$$
Turns ratio $n = \sqrt{(200/5)} = 6.32 : 1$.   (*Ans.*)
(*b*) The current taken from the source $= 2/400$ and the load power P_L is
$$P_L = (2/400)^2 \times 200 = 5 \, \text{mW}.   (Ans.)$$

Power transfer and efficiency

If, in order to obtain maximum power transfer, a load resistance R_L is made equal to the source resistance the efficiency of the system will only be 50% since one-half of the power in the circuit will then be dissipated in the source resistance. The electrical appliances used in the home are not therefore designed with maximum power transfer in mind; instead, in order to gain the maximum possible efficiency the internal resistance of the mains supply is very small, certainly less than 1 Ω. Maximum power transfer is important in applications where the power levels are fairly low and the efficiency is not important. Many examples exist in electronic and radio engineering amongst which are the coupling of an aerial to the RF stage of a radio receiver.

4 D.C. transients

When a d.c. voltage is suddenly applied across a pure resistor the current will suddenly increase from zero to the value predicted by Ohm's law and it will remain at this value for as long as the applied voltage is maintained. When a d.c. voltage is applied to a resistor that is in series with either a capacitor or an inductor this will not be so; instead there will be an initial period during which the current, and the voltages across the two components, change their values followed by the *steady-state* condition when the current and the voltages in the circuit remain constant at particular values. The initially varying current and voltages are known as *transients*.

In the case of the series resistor-capacitor circuit the current initially jumps to the value determined by the resistance, i.e. $I = V/R$, but thereafter it falls exponentially to zero. Conversely, when a d.c. voltage is suddenly applied to a series resistor−inductor circuit the initial current is zero but then it increases exponentially until it reaches its maximum value of $I = V/R$. The three cases are illustrated by Fig. 4.1. Similarly, when the applied voltage is suddenly removed the current in a pure resistor will immediately fall to zero but this is not true for either of the two series circuits. In the resistor−capacitor circuit the capacitor will start to discharge to provide an exponentially decaying current that flows in the opposite direction to the original charging current. When the applied voltage is removed from the inductive circuit a back e.m.f. is induced into the inductance and this keeps the current flowing for a short while.

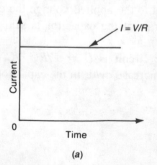
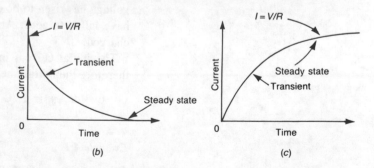

Fig 4.1 Variation of current with time when a d.c. voltage is applied across (*a*) a resistor; (*b*) an *RC* circuit; (*c*) an *LR* circuit

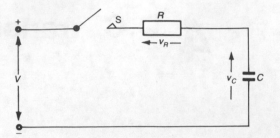

Fig. 4.2 Series *RC* circuit

Resistance—capacitance circuits

Figure 4.2 shows a resistor R connected in series with a capacitor C and an ON/OFF switch S across a d.c. voltage source of V volts. From Kirchhoff's voltage law, the equation for the circuit is

$$V = v_R + v_C \qquad (4.1)$$

where $v_R = iR$ and $v_C = q/C$; i is the instantaneous current flowing in the circuit and q is the instantaneous charge stored in the capacitor.

If the capacitor is initially discharged, so that $q = 0$, the initial voltage across its terminals will be zero and when the switch is closed the initial current that flows into the circuit is limited only by the resistor R. Therefore, the initial current I_I that flows in the circuit is

$$I_I = V/R \qquad (4.2)$$

As this current flows it supplies charge to the capacitor and so a voltage $v_C = qC$ volts appears across the capacitor's terminals. The polarity of this voltage is such that it opposes the applied voltage. The voltage across the resistor is then equal to the applied voltage V minus the voltage v_C across the capacitor, i.e. $V - v_C$, and so the current i in the circuit falls to $i = (V - v_C)/R$. The rate at which charge is supplied to the capacitor also falls and so the capacitor voltage increases less rapidly than before. The further increase in the capacitor voltage means that the current is still further reduced. Hence, as the capacitor charges up and its voltage rises the voltage across the resistor falls and so therefore does the current. Eventually, when the capacitor voltage has risen to be equal to the applied voltage the current will have fallen to zero. At this point the capacitor is said to be fully charged.

The initial current in the circuit is $I_I = V/R = C\, dv_C/dt$ and therefore the initial rate of increase dv/dt in the capacitor voltage is

$$dv_C/dt_{(t=0)} = I_I/C \qquad (4.3)$$

Example 4.1

A d.c. voltage source of 100 V is applied to a 22 μF capacitor that is in series with a 100 kΩ resistor. Determine the initial rate of change of the capacitor voltage.

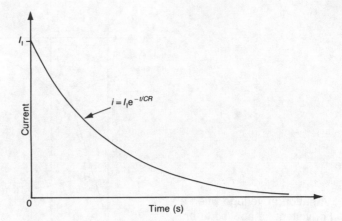

Fig. 4.3 Exponential decay of the current in a series *RC* circuit

Solution

The initial current $I_I = 100/(100 \times 10^3) = 1\,\text{mA}$. Hence

$$\text{d}v_C/\text{d}t = [(1 \times 10^{-3})/(22 \times 10^{-6}) = 45.46\,\text{V/s}. \quad (Ans.)$$

The variation with time of the current flowing in the circuit is exponential as shown by Fig. 4.3. The expression for the current is given by

$$i = I_I\text{e}^{-t/CR} \tag{4.4}$$

The product *CR* is known as the *time constant* of the circuit (see Appendix A).

Time constant

The time constant of an $R{-}C$ circuit is the time in which the current flowing in the circuit would fall from its initial value of $I_I = V/R$ to zero if its initial rate of decrease were to be maintained. This is shown by Fig. 4.4. Alternatively, the time constant is the time that

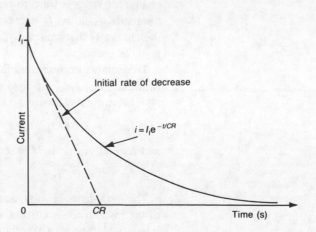

Fig. 4.4 Time constant

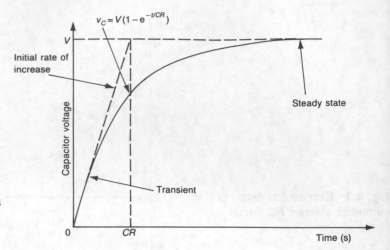

Fig. 4.5 Variation of voltage when a constant d.c. voltage is applied to an *RC* circuit

would be taken for the capacitor voltage to increase from zero to the applied voltage V if the original rate of increase were to be maintained. This is shown by Fig. 4.5; since the capacitor voltage increases exponentially with increase in time it is known as an *exponential growth curve*.

From equation (4.3), at time $t = 0$, $I_I = V/R = C \, dv_C/dt$, so

$$dv_C/dt_{(t=0)} = V/CR \text{ volts/second.}$$

If this rate of change of capacitor voltage were maintained the time t it would take for the capacitor voltage to reach the applied voltage V would be

$$t = V/dv_C/dt = V/(V/CR) = CR \text{ seconds}$$

Thus, the time constant of a $C-R$ circuit is

$$\text{time constant} = CR \text{ seconds} \tag{4.5}$$

This result can be obtained in another way: if the initial rate of change of voltage were to be maintained the charge transferred in t seconds would be $Q = it$ coulombs. To fully charge the capacitor the charge Q that must be supplied is $Q = CV = Vt/R$, and hence $t = CR$ seconds.

The instantaneous current flowing in the circuit is given by equation (4.4), i.e. $i = I_I e^{-t/CR}$. Two times are of particular interest: firstly, $t = CR$ seconds; then

$$i = I_I e^{-1} = 0.368 I_I$$

and secondly, $t = 0.7CR$ seconds when

$$i = I_I e^{-0.7} = 0.5 I_I$$

The instantaneous voltage across the resistor is equal to the product of the instantaneous current and the resistance, i.e. $v = iR$, and so it varies with time in exactly the same way as does the current. The voltage across the capacitor is equal to the applied voltage V minus the voltage v_R dropped across the resistor and so (see Appendix B)

$$v_C = V - iR\,e^{-t/CR} = V - V\,e^{-t/CR}$$

or

$$v_C = V(1 - e^{-t/CR}) \tag{4.6}$$

At time $t = CR$ seconds

$$v_C = V(1 - 0.368) = 0.632\,V$$

and at time $t = 0.7CR$ seconds

$$v_C = V(1 - 0.5) = 0.5\,V$$

Plotting transient curves

The curves showing how the current and/or the capacitor voltage in an $R-C$ circuit vary with time when a d.c. voltage source is suddenly applied can be plotted by calculating values using equations (4.4) and (4.6) respectively provided a calculator with the e^{-x} facility is available.

Example 4.2

A 100 kΩ resistor is connected in series with a 22 μF capacitor. Calculate the time constant of the circuit. If a 24 V d.c. voltage source is applied across the circuit, calculate and plot the variation with time of (a) the circuit current and (b) the capacitor voltage.

Solution
The time constant $CR = 22 \times 10^{-6} \times 100 \times 10^3 = 2.2$ s. (*Ans.*)
(a) The initial current $I_I = 24/(100 \times 10^3) = 240\,\mu$A. The instantaneous current at any time t is hence equal to $i = 240\,e^{-t/2.2}\,\mu$A.
Values of i for various values of time t have been calculated and these are given in Table 4.1. These figures are shown plotted in Fig. 4.6(a). The current will never actually fall to zero but for all practical purposes it is normally taken as being zero after a time equal to *five* time constants, i.e in this particular case after $5 \times 2.2 = 11$ s.
(b) The voltage across the capacitor's terminals is given by
$$v_C = 24\,(1 - e^{-t/2.2})\text{ volts}$$
and for various values of time t has been calculated and is shown in Table 4.2. These figures are shown plotted in Fig. 4.6(b). It can be seen that the capacitor voltage never quite becomes equal to the applied voltage of 24 V but after a time equal to five constants, i.e. 11 s, it is near enough to it for most practical purposes.

Table 4.1

Time t(s)	0	1	2	3	4	5	6	7	8	9	10	11	12	13	14
Current $i(\mu$A)	240	152	97	61	39	25	20	10	6	4	2.6	1.6	1	0.7	0.4

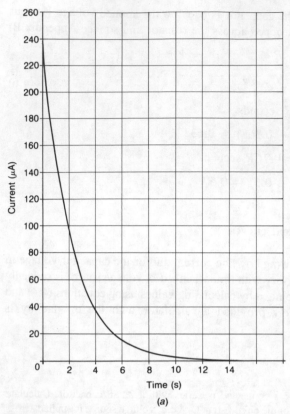

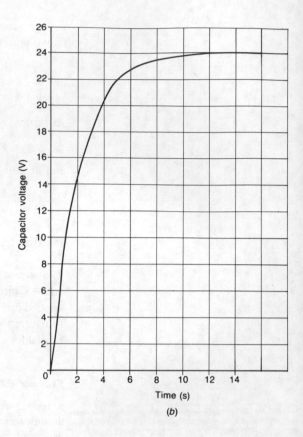

Fig. 4.6

Table 4.2

Time t(s)	0	1	2	3	4	5	6	7	8	9	10	11	12	13	14
Capacitor voltage (V)	0	8.8	14.3	17.9	20.1	22	22.7	23.2	23.5	23.7	23.8	23.87	23.89	23.94	23.96

Approximate method

The calculation of the current flowing in the circuit and the voltage across the capacitor at various times is simple enough if a suitable calculator is available but rather tedious otherwise. Approximate graphs for both the circuit current and the capacitor voltage can be easily and rapidly drawn if it is noted that in each time interval equal to the time constant, the current falls by $e^{-1} = 0.368$ times, and the capacitor voltage rises by $(1 - e^{-1}) = 0.632$ times, its initial value at the beginning of the time interval.

Suppose that the applied voltage is 24 V and the circuit's resistance is 100 kΩ, as in example 4.2, so that the initial current flowing in the circuit is $24/(100 \times 10^3) = 240\,\mu\text{A}$. Figure 4.7(a) shows the variation with time of the circuit current. In the first CR seconds the

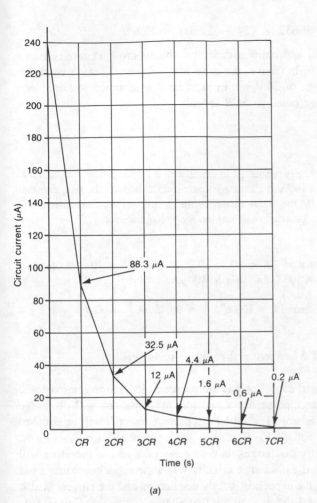

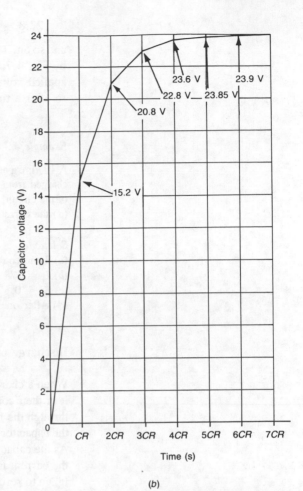

Fig. 4.7

current falls to $0.368 \times 240 = 88.3\,\mu\text{A}$, in the next CR seconds the current falls to $0.368 \times 88.3 = 32.5\,\mu\text{A}$, in the next CR seconds to $0.368 \times 32.5 = 12\,\mu\text{A}$, in the next CR seconds to $0.368 \times 12 = 4.4\,\mu\text{A}$, and after $4CR$ seconds to $0.368 \times 4.4 = 1.6\,\mu\text{A}$, and so on. If straight lines are drawn to join these points together an approximately correct curve will be obtained and this is shown by Fig. 4.7(*a*). Comparing this with the curve given in Fig. 4.6(*a*) it can be seen that the error is fairly small and could always be reduced by 'rounding off' the straight line in Fig. 4.7(*a*), particularly at the higher values of current.

The capacitor voltage growth curve can be similarly obtained. In the first CR seconds the capacitor voltage rises from zero to $0.632 \times 24 = 15.2\,\text{V}$, in the next CR seconds it rises by $0.632 \times (24 - 15.2) = 5.6\,\text{V}$ to $15.2 + 5.6 = 20.8\,\text{V}$, after $3CR$ seconds the capacitor voltage has risen to

$$20.8 + [0.632 \times (24 - 20.8)] = 22.8\,\text{V}$$

and after $4CR$ seconds the capacitor voltage is

$$22.8 + [0.632 \times (24 - 22.8)] = 23.6\,\text{V}$$

and so on. The approximate capacitor voltage growth curve is shown by Fig. 4.7(b). In theory, the capacitor voltage will never reach the applied voltage of 24 V but in practice it is assumed to have done so after a time equal to $5CR$ seconds.

Example 4.3

A 10 μF capacitor is connected in series with a 1 MΩ resistor and then it is charged from a 100 V d.c. voltage source. Calculate (a) the time constant of the circuit, (b) the initial current in the circuit, (c) the final current and (d) the current 1 s after switching on the voltage source.

Solution
(a) Time constant $= CR = 10 \times 10^{-6} \times 1 \times 10^{6} = 10\,\text{s}$. (*Ans.*)
(b) $I_I = V/R = 100/(1 \times 10^{6}) = 100\,\mu\text{A}$. (*Ans.*)
(c) $I_F = 0$. (*Ans.*)
(d) After one second $i = 100\,\text{e}^{-1/10} = 90.48\,\mu\text{A}$. (*Ans.*)

Discharge of a capacitor through a resistor

When a charged capacitor has the source of its charge removed and it is then connected across a resistor, the capacitor will discharge through the resistor. The initial discharge current I_I will be equal to the capacitor voltage V_C divided by the resistance R, i.e. $I_I = V_C/R$. As the capacitor discharges its voltage will fall and so therefore will the current in the discharge circuit. When the capacitor voltage has fallen to zero the capacitor is fully discharged and the current is also zero. Figure 4.8 shows a circuit that allows a capacitor C to be first charged via resistor R and then, when the switch S is operated, to discharge through resistor R. At any time t seconds after the switch is operated to its discharge position the instantaneous current i flowing in the circuit is

$$i = I_I\,\text{e}^{-t/CR} \tag{4.7}$$

where $I_I = $ V/R (see Appendix B).

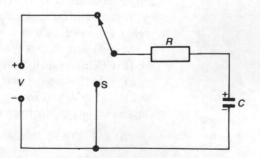

Fig. 4.8 Charge/discharge RC circuit

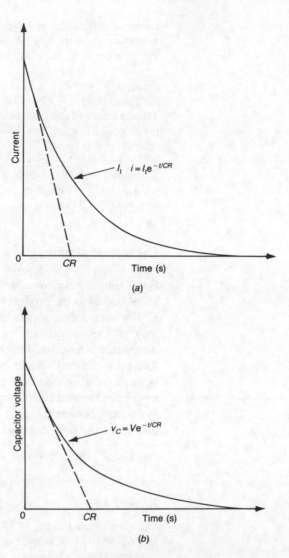

Fig. 4.9 *RC* circuit: decay current,
(*a*) current, (*b*) voltage

The instantaneous capacitor voltage v_C is given by

$$v_C = V\,e^{-t/CR}\ \text{volts} \tag{4.8}$$

The shapes of the current/time and the capacitor voltage/time decay curves are shown by Figs 4.9(*a*) and (*b*) respectively. Now the time constant is the time in which the current, and the capacitor voltage, would take to fall from their initial value to zero if the initial rate of fall were to be maintained. The time constant is again equal to *CR* seconds and it is indicated on both Figs 4.9(*a*) and (*b*).

Example 4.4

In the circuit shown in Fig. 4.8, $V = 6\,\text{V}$, $R = 56\,\text{k}\Omega$ and $C = 0.01\,\mu\text{F}$. If the capacitor is initially discharged, calculate (*a*) the time taken for the capacitor to fully charge, (b) the energy stored in the capacitor when it is fully charged, (c) the time taken for the capacitor voltage to reach 3 V when the capacitor is (i) charging and (ii) discharging.

Solution
(a) Time constant $= CR = 0.01 \times 10^{-6} \times 56 \times 10^3 = 560\,\mu s$.
The capacitor is fully charged after $5 \times 560\,\mu s = 2.8$ ms.   *(Ans.)*
(b) $W = (0.01 \times 10^{-6} \times 36)/2 = 180$ nJ.   *(Ans.)*
(c) (i) $3 = 6\,(1 - e^{-t/560})$
$0.5 = 1 - e^{-t/560}$, $0.5 = e^{-t/560}$, and $\log_e(0.5) = -0.693 = -t/560$.
Therefore $t = 0.693 \times 560 = 388\,\mu s$.   *(Ans.)*
(ii) $3 = 6\,e^{-t/560}$, $\log_e(0.5) = -0.693 = -t/560$,
or $t = 388\,\mu s$.   *(Ans.)*

Calculation of time constant in more complex circuits

For a simple series $R-C$ circuit the time constant of the circuit is simply equal to the product of the resistance R and the capacitance C. For some other, somewhat more complex, circuits, however, it may not be immediately obvious how the time constant can be determined. Consider, for example, the circuit given in Fig. 4.10. The time constant should be calculated by determining the total resistance that is connected across the terminals of the capacitor. In this circuit the total resistance connected across the capacitor is equal to the two resistors R_1 and R_2 connected in parallel with one another and hence the time constant of the circuit is (see Appendix C)

$$(CR_1R_2)/(R_1 + R_2) \tag{4.9}$$

Example 4.5

The circuit shown in Fig. 4.10 has $V = 12$ V, $R_1 = 47$ kΩ, $R_2 = 56$ kΩ and $C_1 = 10\,\mu F$. Calculate (a) its time constant for (i) charge and (ii) discharge, (b) the initial current and (c) the final current taken from the supply.

Solution
(a) (i) The effective resistance of R_1 and R_2 in parallel is
$(47 \times 56)/(47 + 56) = 25.55$ kΩ.
The time constant $= 10 \times 10^{-6} \times 25.55 \times 10^3 = 0.256$ s.   *(Ans.)*

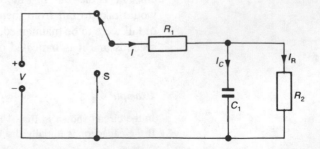

Fig. 4.10 More complex *RC* circuit

(ii) The discharge time constant is the same as the charge time constant, i.e. 0.256 s. (*Ans.*)

(*b*) Initial current = $12/(47 \times 10^3)$ = 255.3 μA. (*Ans.*)

(*c*) Final current = $12/[(47 + 56) \times 10^3]$ = 116.5 μA. (*Ans.*)

It should be noted that the time constants for charge and discharge may not be the same if the circuit is altered in some way after the capacitor has been charged.

Example 4.6

The capacitor in Fig. 4.11 is initially discharged and the applied voltage V is 60 V. When the switch is closed the initial current that flows is equal to 1 mA. 4 ms later the current has fallen to 0.63 mA and after 20 ms it has reached its steady-state value of 0.4 mA. Calculate (*a*) the time constant of the circuit when charging, (*b*) the values of R_1, R_2 and C, (*c*) the energy stored in the circuit when the capacitor is fully charged, (*d*) the time constant of the circuit when it is discharging and (*e*) the capacitor voltage 2 ms after the switch is opened.

Solution

(*a*) $0.63 = 1\ e^{-4/CR}$, $\log_e(0.63) = -0.462 = -4/CR$. Therefore,

time constant $CR = 4/0.462 = 8.66$ ms. (*Ans.*)

(*b*) Initially the capacitor is discharged and so it effectively acts as a short-circuit across resistor R_2. Therefore

$R_1 = 60/(1 \times 10^{-3}) = 60$ kΩ. (*Ans.*)

In the steady state the capacitor is fully charged and the current I_C in it is zero and so the capacitor effectively acts like an open circuit across R_1. The total resistance of the circuit is equal to $60/(0.4 \times 10^{-3}) = 150$ kΩ. Hence

$R_2 = 150 - 60 = 90$ kΩ. (*Ans.*)

Now $(R_1 R_2)/(R_1 + R_2) = (150 \times 90)/(150 + 90) = 56.25$ kΩ. Hence

$C = (8.66 \times 10^{-3})/(56.25 \times 10^3) = 154$ nF. (*Ans.*)

(*c*) In the steady state $V_C = (60 \times 90)/(150 + 90) = 22.5$ V.

$W = CV^2/2 = 0.5 \times 154 \times 10^{-9} \times 22.5^2 = 39$ μJ. (*Ans.*)

(*d*) When discharging only R_2 is in circuit and hence the time constant is

$154 \times 10^{-9} \times 90 \times 10^3 = 13.9$ ms. (*Ans.*)

(*e*) $v_C = 22.5\ e^{-2/13.9} = 19.49$ V. (*Ans.*)

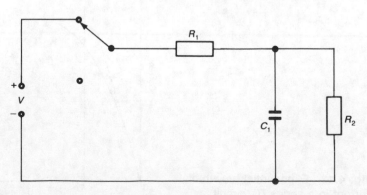

Fig. 4.11

Inductance–resistance circuits

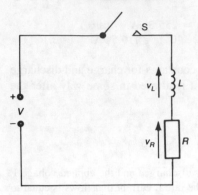

Fig. 4.12 Series *RL* circuit

Figure 4.12 shows an inductor L connected in series with a resistor R and a switch S across a d.c. voltage source of V volts. With the switch open the current in the circuit is, of course, equal to zero. When the switch is closed the current will not instantaneously rise to the value predicted by Ohm's law because of the back e.m.f. that will be generated in the inductance. As soon as a current starts to flow the rate of change of current di/dt is large and so an e.m.f. $e = -L\,di/dt$ is induced into the inductor and, according to Lenz's law, this opposes the applied voltage. Applying Kirchhoff's voltage law to the circuit:

$$V = v_L + v_R = L\,di/dt + iR \tag{4.10}$$

Initially, at time $t = 0$, the current in the circuit is zero and hence $V = L\,di/dt$ and $di/dt = V/L$ amperes/second. This means that the initial rate of change of current in the circuit is equal to the applied voltage divided by the inductance. A short time later the rate of change of current will have fallen so that the applied voltage V has become larger than the back e.m.f. and then a current i flows in the circuit. There will now be a voltage drop iR across the resistor and this will cause the rate of change of current to fall still further, giving another increase in the current and so on. In this way the current in the circuit increases exponentially until, eventually, the final steady-state current, that is limited only by the resistance R in the circuit, i.e $I_F = V/R$, is obtained. The varying current that exists between the time of switching on the voltage supply and the flow of the steady-state current is known as a *transient current*.

The instantaneous current i flowing in an $L–R$ circuit is given by (see Appendix D)

$$i = I_F(1 - e^{-tR/L}) \tag{4.11}$$

where I_F is the steady-state current $= V/R$, and L/R is the time constant of the circuit.

Figure 4.13 shows how the current in an $L–R$ circuit varies with

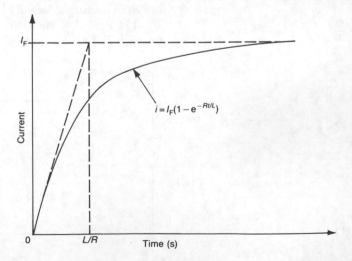

Fig. 4.13 Exponential growth of current in a series *RL* circuit

time after the voltage source has been applied. This is known as an exponential growth curve.

Time constant

The time constant of an $L-R$ circuit is the time in which the current would reach its final steady-state value if its original rate of increase were to be maintained. The time constant is indicated in Fig. 4.13. The initial rate of change of current is V/L amps/second and if this were to be maintained the steady-state current of $I_F = V/R$ would be reached in t seconds, where

$$t = I_F/(di/dt) = (V/R)/(V/L) = L/R \text{ seconds}$$

Thus, the time constant of a series $R-L$ circuit is

$$\text{Time constant} = L/R \qquad (4.12)$$

In practice, the initial rate of increase of the current is not maintained and after a time equal to the time constant the current has only risen to

$$I_F (1 - e^{-1}) = 0.632 I_F$$

Example 4.7

A circuit consists of a 2 H inductor and a 10 Ω resistor connected in series across a 2 V d.c. voltage source. Calculate, and plot, the current in the circuit for the first one second after the voltage source is applied to the circuit. Use the approximate method of calculation.

Solution
The time constant $L/R = 2/10 = 0.2$ s.
The final steady-state value of the current is $I_F = 2/10 = 0.2$ A.
After 0.2 s $i = 0.632 \times 200 = 126.4$ mA.
After 0.4 s $i = 0.632 \times (200 - 126.4) + 126.4 = 46.5 + 126.4 = 172.9$ mA.
After 0.6 s $i = 0.632 \times (200 - 172.9) + 172.9 = 17.1 + 172.9 = 190$ mA.
After 0.8 s $i = 0.632 \times (200 - 190) + 190 = 6.3 + 190 = 196.3$ mA.
After 1 s $i = 0.632 \times (200 - 196.3) + 196.3 = 2.3 + 196.3 = 198.6$ mA.
These calculated values are shown plotted in Fig. 4.14.

Example 4.8

An 800 mH coil of resistance 80 Ω is connected across a 24 V d.c. voltage source. Calculate (a) the final steady-state current, (b) the time constant, (c) the current after 20 ms, and (d) the energy stored in the circuit once the current has reached its steady-state value.

Solution
(a) $I_F = V/R = 24/80 = 0.3$ A. *(Ans.)*
(b) Time constant $= L/R = 0.8/80 = 0.01$ s $= 10$ ms. *(Ans.)*

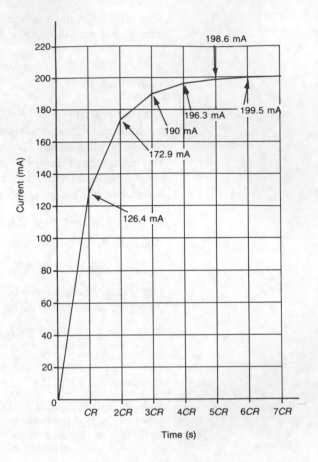

Fig. 4.14

(c) $i = 300 \, (1 - e^{-20/10}) = 259.4$ mA.   (*Ans.*)
(d) $W = LI^2/2 = 0.5 \times 0.3^2 \times 0.8 = 36$ mJ.   (*Ans.*)

Decay of current

While the steady-state current is flowing in the inductance a magnetic field is set up around the inductance. When the current is suddenly turned off the current will fall and so the magnetic field will collapse. There is then a change in the flux linkages and this change causes a back e.m.f. $v_L = -L \, di/dt$ volts to be induced into the inductance. By Lenz's law the polarity of this induced e.m.f. is such that it tends to keep the current flowing; at time $t = 0$ the current is limited only by the resistance R and hence $v_L = v_R$. As the rate of change of current decreases v_L falls and the current falls exponentially towards zero. Since $v_R = iR$ the voltage dropped across the resistance R also decays exponentially.

The instantaneous current i is given by (see Appendix E)

$$i = I_I \, e^{-tR/L} \tag{4.13}$$

The shape of the current/time curve of the decay current is shown

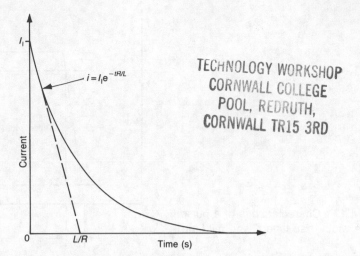

Fig. 4.15 Exponential decay of current in a series *RL* circuit

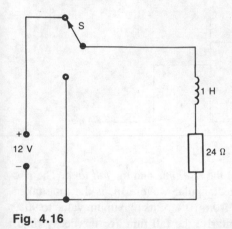

Fig. 4.16

by Fig. 4.15; as before, the time constant is time in which the current would fall to zero if the initial rate of fall were to be maintained and this time is marked on the graph.

Example 4.9

For the circuit given in Fig. 4.16 calculate (*a*) the current flowing in the circuit immediately before the switch is changed over, (*b*) the time constant of the circuit and (*c*) the time that elapses before the current flowing in the circuit has fallen to (i) 100 mA and (ii) 10 mA.

Solution

(*a*) $I_I = 12/24 = 0.5$ A.   (*Ans.*)

(*b*) Time constant $= L/R = 1/24$ s $= 41.67$ ms.   (*Ans.*)

(*c*) (i) $i = 100 = 500\,e^{-24t}$, $\log_e 0.2 = -1.61 = -24t$, or
$t = 67.1$ ms.   (*Ans.*)

(ii) $i = 10 = 500\,e^{-24t}$, $\log_e 0.02 = -3.91 = -24t$, or
$t = 163$ ms.   (*Ans.*)

The effect of circuit time constant on rectangular waveforms

When a rectangular waveform is passed through a circuit that contains both resistance and capacitance, or both resistance and inductance, it may, or it may not, have its waveform changed. The determining factor is the relative values of the circuit's time constant and the periodic time of the pulse waveform. Any change in the rectangular waveform that occurs may, or may not, be one that is required depending upon the application.

Characteristics of pulse waveforms

All rectangular pulses take a non-zero time to change from their minimum value to their maximum value or vice versa. The time taken

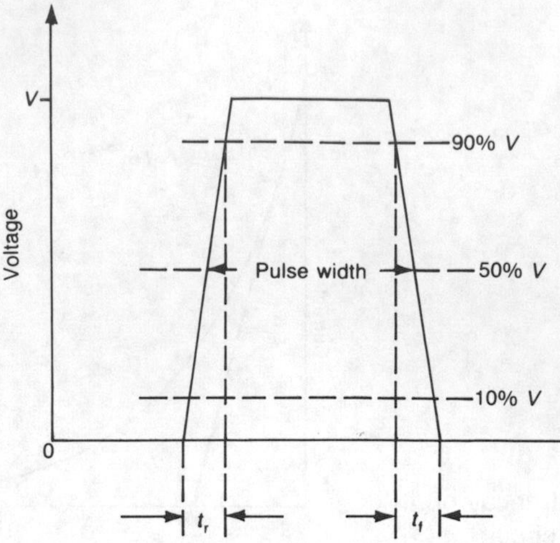

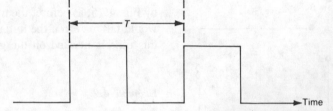

Fig. 4.17 Characteristics of a pulse: pulse width, rise time t_r and fall time t_f

Fig. 4.18 Pulse repetition frequency = $1/T$

is usually quoted in terms of the *rise time* and the *fall time*. The rise time t_r of a pulse, or of a rectangular waveform, is the time taken for its amplitude to increase from 10% of its maximum value to 90% of its maximum value. Similarly, the fall time (or the decay time) t_f of a pulse is the time taken for the amplitude of the pulse to fall from 90 to 10% of its maximum value. The *width* or *duration* of a pulse is taken as the distance along the time axis between the 50% amplitude points on the leading and trailing edges of the pulse. These three terms are illustrated by Fig. 4.17. The *pulse repetition frequency* (PRF) is the number of pulses that occur in one second and it is equal to the reciprocal of the periodic time T of the pulse waveform. The periodic time is the time that elapses between the leading edges of two consecutive pulses as shown by Fig. 4.18.

Rise time

When the amplitude of a pulse is equal to 10% of its final value,

$$0.1V = V(1 - e^{-t/CR})$$
$$0.9 = e^{-t/CR}$$
$$1/0.9 = 1.11 = e^{t/CR}$$

and

$$t_1/CR = \log_e 1.11 = 0.1 \quad \text{or} \quad t_1 = 0.1CR$$

When the pulse amplitude is equal to 90% of its final amplitude,

$$0.9V = V(1 - e^{-t/CR})$$
$$0.1 = e^{-t/CR}$$
$$1/0.1 = 10 = e^{t/CR}$$

and

$$t_2/CR = \log_e 10 = 2.3 \quad \text{or} \quad t_2 = 2.3CR.$$

Therefore, the rise time $t_r = t_2 - t_1$ or

$$t_r = 2.2CR \text{ seconds} \tag{4.14}$$

Rectangular pulses applied to an $R-C$ circuit

Figure 4.19 shows a perfectly rectangular pulse of width τ seconds and amplitude V volts applied to an $R-C$ circuit. The output voltage of the circuit is the voltage across the resistor. Suppose that the capacitor is initially discharged and that the time constant of the circuit is much less than the pulse width. When the first pulse is applied to the circuit the discharged capacitor acts like a short-circuit and so the output voltage instantaneously changes from 0 V to V volts. The flow of current will charge the capacitor up, with the polarity shown in the figure, and so its voltage will rise. Then, since $V = v_C + v_R$, the output voltage will fall. Since the pulse width is much less than the time constant of the circuit there will only have been a small reduction in the output voltage when the pulse ends.

At the end of the pulse the capacitor will discharge through the resistor R but the discharge current is in the opposite direction to the charging current. This means that the output voltage will suddenly fall by $-V$ volts and this fall takes the output voltage slightly negative before it decays exponentially towards 0 V. The output voltage pulse is shown by Fig. 4.20(a). If the time constant of the circuit is reduced relative to the pulse width the output voltage will exhibit a much greater variation and this is illustrated by Figs 4.20(b) and (c). Clearly, if an $R-C$ circuit is required to pass a rectangular pulse waveform with little waveform distortion it is necessary for the time constant of the circuit to be much longer than the pulse width.

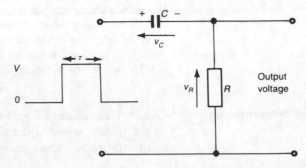

Fig. 4.19 Rectangular pulse applied to an *RC* circuit

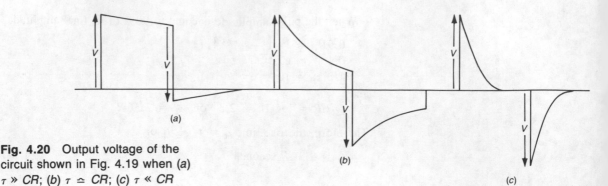

Fig. 4.20 Output voltage of the circuit shown in Fig. 4.19 when (a) $\tau \gg CR$; (b) $\tau \simeq CR$; (c) $\tau \ll CR$

Example 4.10

A 10 V rectangular pulse of width 1 ms is applied to an $R-C$ circuit whose time constant is 2 ms. Determine the waveform of the output voltage for four successive pulses if the output voltage is taken from (a) the resistor and (b) the capacitor.

Solution
(a) *Pulse 1*: Initial output voltage = 10 V.
Output voltage at end of pulse = $10 \, e^{-1/2}$ = 6.1 V.
Negative voltage = 6.1 − 10 = −3.9 V.
Negative voltage at start of next pulse = $-3.9 \, e^{-1/2}$ = −2.37 V.
Pulse 2: Output voltage = −2.37 + 10 = 7.63 V.
Output voltage at end of pulse = $7.63 \, e^{-1/2}$ = 4.63 V.
Negative voltage = 4.63 − 10 = −5.37 V.
Negative voltage at start of next pulse = $-5.37 \, e^{-1/2}$ = −3.26 V.
Pulse 3: Output voltage = −3.44 + 10 = 6.74 V.
Output voltage at end of pulse = $6.74 \, e^{-1/2}$ = 4.09 V.
Negative voltage = 4.09 − 10 = −5.91 V.
Negative voltage at start of next pulse = $-5.91 \, e^{-1/2}$ = −3.59 V.
Pulse 4: Output voltage = −3.59 + 10 = 6.41 V.
Output voltage at end of pulse = $6.41 \, e^{-1/2}$ = 3.89 V.
Negative voltage = 3.89 − 10 = −6.11 V.
Negative voltage at start of next pulse = $-6.11 \, e^{-1/2}$ = −3.71 V.
Further values have been similarly calculated for pulses five, six, etc. and it can be seen from Fig. 4.21(a) that the output voltage has a steady-state value that varies between ±6.2 and ±3.8 V.
(b) Since $V = v_C + v_R$ the output voltage taken from the capacitor is equal to 10 − v_R. Hence, the output voltage is as shown by Fig. 4.21(b). This means that after several pulses have been applied to an $R-C$ circuit the voltage across the resistor becomes symmetrically positioned either side of the zero voltage axis. The voltage across the capacitor is then of approximately triangular shape and is centred about a mean value equal to (6.2 + 3.8)/2 = 5 V or one-half of the applied voltage.

Integrating and differentiating circuits

An $R-C$ circuit can be used either to integrate or to differentiate an input voltage provided the choice of the circuit's time constant is correctly made. When an $R-C$ circuit is to be employed as an

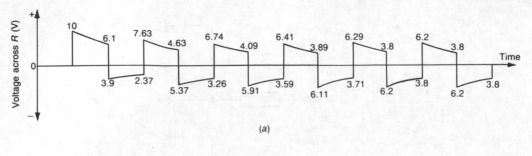

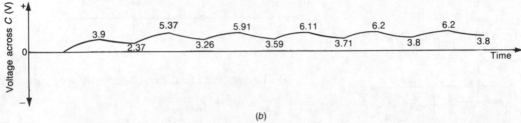

Fig. 4.21

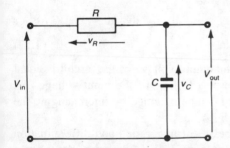

Fig. 4.22 Integrating circuit

integrator the output voltage is taken from the capacitor and the time constant of the circuit must be long relative to the pulse width τ. Conversely, if the circuit is to be employed as a differentiator the output voltage is taken from the resistor and the circuit's time constant should be much shorter than the pulse width τ.

Figure 4.22 shows a basic integrating circuit. From this

$$V_{in} = v_C + v_R$$

but since $CR \gg \tau$,

$$v_C \ll v_R \quad \text{and so} \quad V_{in} \simeq v_R$$

The output voltage is

$$
\begin{aligned}
V_{out} = v_C &= Q/C \\
&= (1/C) \int i \, dt = (1/C) \int (v_R)/R) \, dt \\
&= (1/C) \int (V_{in}/R) \, dt
\end{aligned}
$$

or

$$V_{out} = (1/CR) \int V_{in} \, dt \tag{4.15}$$

Equation (4.15) states that the output voltage of the circuit is equal to the integral with respect to time of the input voltage.

A more accurate integration of a waveform can be obtained if an op-amp integrator is employed and Fig. 4.23 shows the basic circuit. The inverting terminal is a virtual earth point and hence an input current $I_{in} = V_{in}/R_1$ flows through R_1 and then, since the input impedance of the op-amp is very high, through C_1. Therefore

$$
\begin{aligned}
V_{out} &= -(1/C_1) \int i \, dt \\
&= -(1/C_1) \int (V_{in}/R_1) \, dt \\
&= -(1/C_1 R_1) \int V_{in} \, dt
\end{aligned}
\tag{4.16}
$$

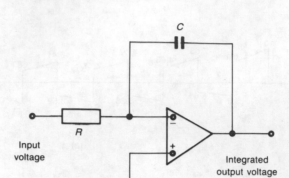

Fig. 4.23 Op-amp integrator

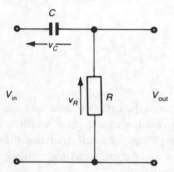

Fig. 4.24 Differentiating circuit

The basic circuit of a differentiator is shown by Fig. 4.24. From this

$$V_{in} = v_C + v_R$$

but since $CR \leq \ll \tau$,

$$v_R \leq v_C \quad \text{and so} \quad V_{in} \simeq v_C$$

The current i in the circuit is $i = \mathrm{d}q/\mathrm{d}t = C \, \mathrm{d}v_C/\mathrm{d}t = C \, \mathrm{d}V_{in}/\mathrm{d}t$, or

$$V_{out} = iR = CR \, \mathrm{d}V_{in}/\mathrm{d}t. \tag{4.17}$$

This equation states that the output voltage of the circuit is equal to the differentiation with respect to time of the input voltage. An op-amp differentiating circuit can be obtained by interchanging the positions of R_1 and C_1 in Fig. 4.23.

As an example of the processes of integration and differentiation when applied to a signal waveform, Fig. 4.25 shows that when a

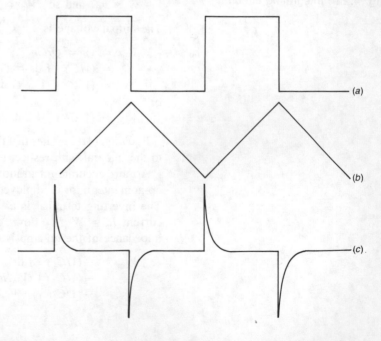

Fig. 4.25 Integration and differentiation of a square waveform

square waveform is applied to an integrating circuit the output voltage is of triangular waveform, and when the same waveform is applied to a differentiating circuit the output voltage consists of a series of alternately positive and negative narrow pulses.

Appendix A   Voltage growth and current decay in an $R-C$ circuit

When a constant voltage is applied across an $R-C$ circuit the current t seconds later will be i amperes and the capacitor voltage will be v_C volts. If the capacitor voltage changes by dv_C volts in dt seconds the current i will be $i = C\,dv_C/dt$ amperes and the voltage v_R across the resistor will be

$$v_R = iR = CR\,dv_C/dt \text{ volts}$$

Now

$$V = v_C + v_R = v_C + CR\,dv_C/dt$$
$$V - v_C = CR\,dv_C/dt$$
$$dt/CR = dv_C/(V - v_C).$$

Integrating both sides gives

$$t/CR = -\log_e(V - v_C) + K$$

When time $t = 0$, $v_C = 0$ and therefore $K = \log_e V$. Hence

$$t/CR = -\log_e(V - v_C) = \log_e V.$$
$$e^{t/CR} = V/(V - v_C)$$
$$V\,e^{t/CR} - v_C\,e^{t/CR} = V$$
$$v_C\,e^{t/CR} = V\,(e^{t/CR} - 1).$$

And

$$v_C = V\,(1 - e^{-t/CR}) \text{ V.} \tag{4.18}$$

Also,

$$i = C\,dv_C/dt = CV\,d(1 - e^{-t/CR})/dt$$
$$= CV\,(1/CR)\,e^{-t/CR}$$

or

$$i = (V/R)\,e^{-t/CR} \text{ A} \tag{4.19}$$

Appendix B   Voltage and current decay in an $R-C$ circuit

When a fully charged capacitor C is discharged through a resistance R the current t seconds after the start of the discharge is $i = -v_C/R$. If the capacitor voltage changes by dv_C volts in dt seconds

$$i = C\,dv_C/dt$$

Therefore $-v_C/R = C\,dv_C/dt$, and $dt/CR = -v_C/v_C$. Integrating both sides of this equation gives

$$t/CR = -\log_e v_C + K$$

When time $t = 0$, $v_C = V$ and hence $K = \log_e V$. Hence

$$t/CR = -\log_e v_C + \log_e V = \log_e(V/v_C).$$
$$V/v_C = e^{t/CR}$$

and

$$v_C = V\,e^{-t/CR} \text{ V.} \tag{4.20}$$

and
$$i = (V/R) \, e^{-t/CR} \text{ A} \tag{4.21}$$

Appendix C   Time constant

Referring to the circuit shown by Fig. 4.11,

$$i = i_C + i_R = (V - v_C)/R_1, \quad I_2 = v_C/R_2 \text{ and } v_C = q/C$$

Therefore,

$$(V - v_C)/R_1 = i_C + v_C/R_2$$
$$V/R_1 - i_C = v_C \, (1/R_1 + 1/R_2)$$
$$= (q/C)[(R_1 + R_2)/(R_1R_2)]$$
$$q = C[R_1R_2/(R_1 + R_2)] \, [V/R_1 - i_C]$$
$$dq/dt = i_C = [-CR_1R_2/(R_1 + R_2)]di_C/dt$$

Thus the current i_C flowing in the capacitor is proportional to its own rate of change and this means that i_C is exponential. Suppose therefore that $i_C = A \, e^{-xt}$. At time $t = 0$, $q = v_C = i_R = 0$, and so $i_C = i = V/R_1$. Hence $A = V/R_1$. Now

$$di_C/dt = -i_C/[CR_1R_2/(R_1 + R_2)]$$

and hence

$$di_C/dt_{(t=0)} = -(V/R_1)/[CR_1R_2/(R_1 + R_2)]$$

The initial value of i_C is V/R_1 and this value falls exponentially to zero. It would reach zero in $CR_1R_2/(R_1 + R_2)$ seconds if the initial rate of decrease were to be maintained. Therefore, the time constant is $CR_1R_2/(R_1 + R_2)$ seconds.

When the switch is thrown R_1 is placed in parallel with R_2 and so the time constant for discharge is again equal to $CR_1R_2/(R_1 + R_2)$ seconds. This means that, provided the circuit has not been changed, the time constants for charge and discharge are equal to one another and it will often prove easier to calculate the time constant by considering the discharge of the capacitor.

Appendix D   Current growth in an inductive circuit

The current t seconds after the voltage source has been applied to an inductance and a resistance in series is i amperes and the rate of increase in the current is di/dt amperes/second. The e.m.f. induced into the inductance is equal to $-L \, di/dt$. Therefore,

$$V = L \, di/dt + iR$$
$$V - iR = L \, di/dt \quad V/R - i = I_F - i = (L/R)di/dt.$$

Hence

$$(R/L)dt = di/(I_F - i).$$

Integrating both sides of this equation gives

$$Rt/L = -\log_e(I_F - i) + K$$

At time $t = 0$, $i = 0$ and so $K = \log_e I_F$. Hence

$$Rt/L = -\log_e(I_F - i) + \log_e I_F = \log_e[I_F/(I_F - i)]$$
$$e^{Rt/L} = I_F/(I_F - i) \quad (I_F - i)/I_F = e^{-Rt/L}$$

and
$$i = I_F (1 - e^{-Rt/L}) \text{ A} \tag{4.22}$$

Appendix E   Current decay in an inductive circuit

Suppose that the steady-state current in an $L-R$ circuit is I_F. When the voltage source is removed and replaced by a short-circuit the current will be i. Then, from Kirchhoff's voltage law,

$$0 = iR + L \, di/dt.$$

Hence
$$iR = -L \, di/dt \quad \text{and} \quad R \, dt/L = -di/i$$

Integrating both sides of this equation gives

$$Rt/L = -\log_e i + K$$

At time $t = 0$, $i = I_I$, and hence

$$0 = \log_e I_I - \log_e i = \log_e(I_I/i)$$
$$I_I/i = e^{Rt/L} \quad \text{and} \quad i = I_I e^{-Rt/L} \text{ A}. \tag{4.23}$$

5 Electrical machines

A rotating electrical machine is able to operate as either a generator or as a motor. The rotational speed N of a machine is usually given in revolutions per second (r.p.s.) or in revolutions per minute (r.p.m.), where 1 r.p.s. $= 2\pi$ rad/s. The generator and motor modes of operation of an electrical machine are illustrated by Figs 5.1(a) and (b) respectively. In the generator mode an applied torque T causes the shaft of the machine to rotate at an angular velocity of ω radians/second. The shaft has an armature winding mounted on it and this winding is moved through a uniform magnetic field and so has an e.m.f. E induced into it. This e.m.f. causes a current I to flow in the external load which has the terminal voltage of V volts across it. The voltage V is always smaller than the induced e.m.f. E. The input power $P_{in} = \omega T$ watts and the output power $P_{out} = VI$ watts. When a machine is used as a motor, a voltage V is applied to its input terminals and this causes a current I to flow into the machine. This current causes a torque T to be exerted that makes the shaft of the machine rotate with a speed of N r.p.s. This rotation generates an internal e.m.f. of E volts that opposes the applied voltage and so reduces the input current.

A generator is an electrical machine which converts mechanical input energy into electrical output energy. Whenever there is a relative motion between a conductor and a magnetic field an e.m.f. will be induced into that conductor. This is known as the *generator effect*. The relative motion may occur because the conductor rotates about an axis through the magnetic field, or because the conductor is kept in a fixed position and the magnetic field is made to move around it. Faraday's law of electromagnetic induction states 'that the magnitude of the induced e.m.f. is equal to the rate of change of the flux linkages', i.e. $E = -d\Phi/dt$ volts. If the length of the conductor

Fig. 5.1 Electrical machine used as (*a*) a generator, and (*b*) a motor

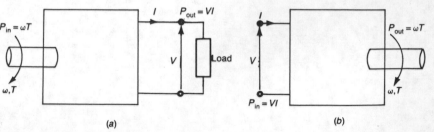

(a)

(b)

in the magnetic field is l metres and the conductor moves perpendicularly to the magnetic field, with an average velocity of v metres/second for t seconds, the distance covered will be $x = vt$ metres. Then the induced e.m.f. E is

$$E = dBA/dt = dBlx/dt = Blv \text{ V}. \tag{5.1}$$

where B is the average flux density in tesla of the magnetic field.

In a generator the conductor follows a circular path through the magnetic field with an angular velocity of ω radians/second. If the radius of the circular path is r metres then the equivalent linear velocity v is $v = \omega r$ metres/second. Therefore

$$E = Bl\omega r \text{ V}. \tag{5.2}$$

The flux density is directly proportional to the magnetic flux Φ, in webers, and hence the induced e.m.f. is

$$E = k_e\Phi\omega \text{ V}. \tag{5.3}$$
or
$$E = k_e N\Phi \text{ V}. \tag{5.4}$$

Thus the induced e.m.f. is directly proportional to the product of the average magnetic flux and the speed of the machine. The constant k_e is known as the *e.m.f. constant* of the generator.

When a current-carrying conductor is situated in a uniform magnetic field a force will be exerted upon that conductor because of the interaction between the uniform field and the field set up by the current. The magnitude of this force is given

$$F = BIl \text{ N}. \tag{5.5}$$

where B is the average flux density of the uniform magnetic field in tesla, I is the current flowing in the conductor in amperes, and l is the length of conductor in the magnetic field.

If the conductor is free to rotate about an axis at its centre it will follow a circular path when this force is applied to it. The flux density of the magnetic field is directly proportional to the flux Φ and hence the torque T applied to the conductor ($T = Fr$, where r is the radius of the circular path) is

$$T = k_t\Phi I \text{ N m}. \tag{5.6}$$

Thus the torque applied to a current-carrying conductor in a magnetic field is directly proportional to the product of the magnetic flux and the current. The constant k_t is known as the *torque constant* of the machine. Equation (5.6) can be rearranged to give $I = T/kt\Phi$ which shows that the current taken by a motor from the voltage supply varies directly with the load torque if the flux is constant. Equations (5.4) and (5.6) apply to both d.c generators and d.c. motors.

The current I_a that flows through the *armature* of a d.c. machine drops a voltage $I_a r_a$ across its armature resistance r_a. This voltage drop results in a small power loss inside the machine. The rotation of the armature within the magnetic field generates a *back e.m.f.* of E volts and hence the terminal voltage V is

for a generator: $V = E - I_a r_a$ (5.7)

for a motor: $V = E + I_a r_a$ (5.8)

Example 5.1

For the d.c. shunt motor shown in Fig. 5.2 calculate the current I taken from the supply.

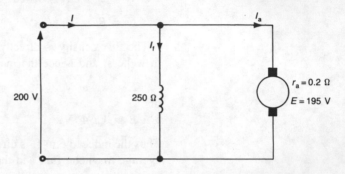

Fig. 5.2

Solution
$I = I_a + I_f = 200/250 + (200 - 195)/0.2 = 25.8$ A.   (*Ans.*)

In a generator the input mechanical power is equal to $T\omega$ watts and the output power is (a) $\sqrt{(3)}V_L I_L \cos\phi$ for a three-phase machine, (b) $VI \cos\phi$ for a single-phase machine or (c) VI watts for a d.c. machine. In the case of a motor one of the three electrical powers is the input power and the mechanical output power is ωT. The output power is always less than the input power because of inevitable losses within the machine and this means that the efficiency

$$\eta = (\text{output power})/(\text{input power}) \times 100\% \qquad (5.9)$$

is less than 100%. The internal power losses are:

(a) copper ($I^2 R$) losses in the armature resistance and in the resistance of the field winding;
(b) iron losses caused by hysteresis and eddy current losses in the magnetic circuit;
(c) mechanical losses caused by friction and by windage (i.e. losses due to the air circulating inside the machine).

Example 5.2

A 300 V d.c. motor takes an input current of 3 A when it produces an output torque of 8 Nm at a speed of 1000 r.p.m. Calculate the efficiency of the motor.

Solution
Input electrical power = $300 \times 3 = 900\,\text{W}$.
Output mechanical power = $(2\pi \times 1000 \times 8)/60 = 837.8\,\text{W}$.
Efficiency = $837.8/900 \times 100 = 93.1\%$ (*Ans.*)

The magnetic circuit of most machines has two parts, a stationary part known as the *stator*, and a rotating part known as the *rotor*. The two parts are made from magnetic materials and are separated from one another by an air gap. In many machines the rotor and/or the stator are laminated to reduce eddy currents. One of the two parts must carry the *field winding* whose function it is to produce the main magnetic field for the machine. The other part, usually called the *armature*, must carry the main current that flows in the machine. Some machines have the field winding on the stator and the armature on the rotor while other machines are arranged the other way around, the rotor carrying the field winding and the stator carrying the armature. (Small motors may employ a permanent magnet to produce the main magnetic field instead of a field winding and these magnets are always part of the stator.) The most common choice of motor when either speed control or torque control is required is the d.c. motor. Increasing use is being made, however, of the d.c./a.c. converter fed induction motor for controlled-speed applications.

The a.c. generator

The essential parts of an a.c. generator are shown by Fig. 5.3. Two pole pieces that have north and south polarities are used to produce a strong, uniform, magnetic field in the space between them. A coil of wire is mounted on a shaft that is free to rotate about a central axis to form the armature of the generator. At the end of the armature

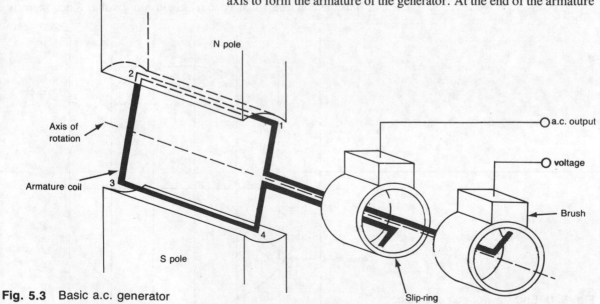

Fig. 5.3 Basic a.c. generator

shaft two slip-rings are mounted and each one has one end (marked as 1 and 4) of the coil connected to it. Each slip-ring has a carbon brush held in sliding contact with it by a spring-loaded brush holder. As the motor shaft rotates because of an applied torque the coil rotates through the magnetic field and this produces a change in the flux linkages. An e.m.f. is therefore induced into the coil and this is supplied to the motor's output terminals by the two brushes.

When the coil is in such a position that its two long sides, 1−2 and 3−4, move in the same direction as the magnetic field there will be no change in the flux linkages and so the instantaneous induced e.m.f. will be zero. As the coil continues to rotate the changes in flux linkages increase and reach their maximum value when the coil position is such that its two long sides are moving perpendicularly to the magnetic field. When the coil passes through this point the rate of change in the flux linkages falls and will reach zero when the coil is 180° from its original position and once again its two long sides are moving in the same direction as the magnetic field.

An end view of the coil is shown in Fig. 5.4. The coil rotates about its axis of rotation, marked as O, with an angular velocity of ω radians/second. The plane of the coil makes an angle θ with the reference horizontal direction. Consider the side 1−2 of the coil; the linear velocity, $v = \omega r$ metres/second, acts tangentially to the circular path through which the coil rotates. This tangential motion can be resolved into its horizontal component $v \sin \theta$ metres/second, and its vertical component $v \cos \theta$ metres/second. The vertical component of the velocity is in line with the magnetic field and so it will not induce an e.m.f. into the conductor 1−2. The horizontal component is at right angles to the magnetic field and so it does induce an e.m.f. E into the conductor that is equal to $Blv \sin \theta$ volts. Since there is

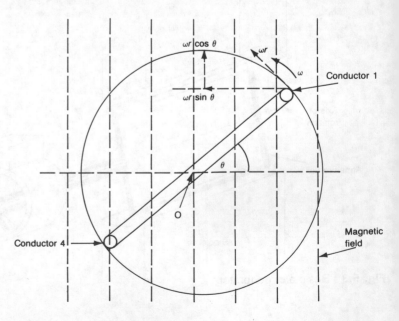

Fig. 5.4 End view of coil in an a.c. generator

an equal, in-phase, voltage induced into the other conductor 3−4 the total induced e.m.f. is $E = 2Blv \sin \theta$ volts. (The two short sides of the coil that join the conductors 1−2 and 3−4 do not contribute to the induced e.m.f. since their induced voltages are in anti-phase and hence cancel one another out.) The a.c. voltage output of the generator varies sinusoidally between peak values of $\pm 2Blv$ volts. If the coil has n turns each turn will have the same voltage induced into it and hence the total induced e.m.f. is $E = 2nBlv \sin \theta$ volts.

The angular velocity of the armature is ω radians/second and if the radius of the coil is r metres the linear velocity of a conductor is $v = \omega r$ metres/second. Therefore the induced e.m.f. is

$$E = 2nBl\omega r \sin \theta \ \text{V}. \tag{5.10}$$

A practical a.c. generator will have several pairs of poles whose polarities are alternately north and south. If the number of *pairs* of poles is p then the induced e.m.f. is

$$E = 2npBl\omega r \ \text{V}. \tag{5.11}$$

It is more usual to think of the speed of the generator in terms of revolutions/minute rather than radians/second. The speed N is equal to $2\pi\omega/60$ r.p.m. and so the expression for the induced e.m.f. can be written as

$$E = 60npBlrN/\pi \ \text{V}. \tag{5.12}$$

The frequency of the a.c. output voltage is equal to the product of the number of pairs of poles p and the speed of the armature in r.p.s.

A d.c. voltage can be obtained by rectifying the a.c. output voltage. Alternatively, the slip-rings can be replaced by a *commutator* to produce a *d.c. generator*.

Motor car alternator

The battery of a motor car is kept in a fully charged state by a three-phase a.c. alternator whose output voltage is rectified by a six-diode rectifier unit. The rotor carries the field windings that are wound on six poles that are alternately of north and south polarity and the laminated stator carries three windings. Alternating voltages are induced into each of the three windings by the rotating magnetic field, and the spacing of the windings is such that a three-phase output voltage is generated. The three-phase output voltage is full-wave rectified by the circuit shown in Fig. 5.5.

The d.c. generator

The alternating output voltage of an a.c. generator can be turned into a unidirectional voltage if the sense of the rotor windings can be reversed at the correct instants in time. The switching will ensure that alternate half-cycles of the generated a.c. voltage have their

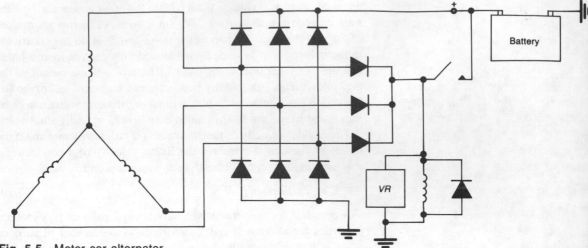

Fig. 5.5 Motor car alternator system

polarity reversed. The necessary switching is performed by a *commutator* which consists of a number of brass segments that are insulated from one another by strips of mica. Connections from the commutator to the output terminals of the generator are made by two stationary carbon brushes; these are mounted in brush holders and held in firm contact with the brass segments by spring pressure. The brushes therefore make a sliding contact with the segments as they pass beneath the brushes.

Figure 5.6 shows the simplest case of a two-segment commutator. The two ends of the coil of wire are each connected to one segment

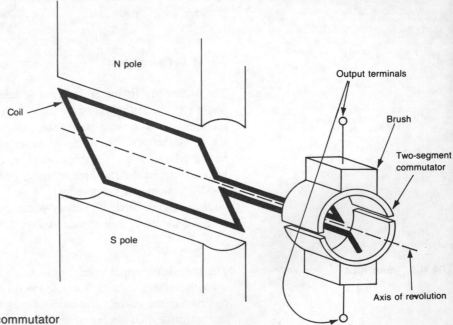

Fig. 5.6 Two-segment commutator

of the commutator. The two brushes supply the generated voltage to the output terminals of the generator. The brushes are positioned so that their electrical contact with the commutator changes from one segment to the other at those instants in time when the generated a.c. voltage is instantaneously zero and just about to change polarity. Thus, in this simple case, the brushes are mounted in the horizontal plane since the induced e.m.f. is zero when the coil is in the vertical position. Thus the action of the commutator is to ensure that the voltage that appears at the output terminals is always of the same polarity, i.e. it is a d.c. voltage.

The output voltage consists of a series of half-sinewave pulses as shown by Fig. 5.7. Clearly, although the output voltage is direct it is far from constant and it would be unsuitable for most purposes. A smoother d.c. output voltage could be obtained if the commutator were divided into four segments, and two armature coils were employed, as shown by Fig. 5.8. Suppose that at the instant in time

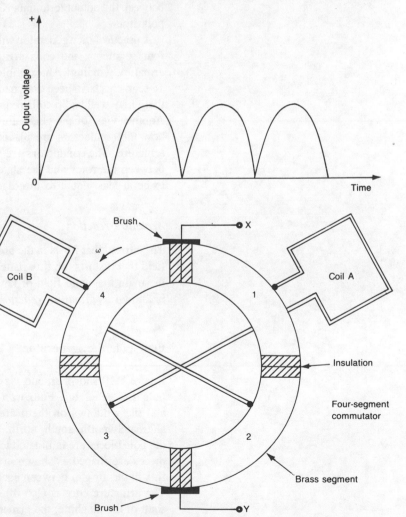

Fig. 5.7 Output voltage of two-segment commutator machine

Fig. 5.8 Four-segment commutator

when the segments 1 and 3 are in contact with the brushes the e.m.f. induced into coil A is positive at the upper brush and negative at the lower brush. The voltage at the output terminal marked X is then positive. A short time later the two coils and the commutator will have rotated until segment 2 is in contact with the upper brush and segment 4 is in contact with the lower brush. Coil B is now in the position that coil A was in originally, and so, once again, the upper brush receives the positive polarity, and the lower brush receives the negative polarity, of the induced e.m.f. A further short time later brings segment 3 in contact with the upper brush and segment 1 in contact with the lower brush. Coil A has now reversed its original position and hence the induced e.m.f. appearing at segment 3 has the opposite polarity to before, i.e. it is now positive. This positive voltage appears at the output terminal marked as X. The induced voltage at segment 1, and hence at the lower brush, is now negative and this voltage appears at terminal Y. This means that the voltage appearing between the output terminals XY of the generator always has the same polarity.

A practical d.c. generator will have many more coils and more than four segments; the coils are connected in series to form a single armature winding. An example of this is shown in Fig. 5.9(*a*); only 16 segments have been drawn to keep the drawing simple but typically, there may well be 36 coils requiring a commutator with 72 segments. Another view of the generator is given by Fig. 5.9(*b*) which shows how the conductors are placed in slots cut along the length of the armature. The conductors are placed in slots to minimize the air gap between the rotor and the stator. Otherwise the air gap would have to be at least equal to the diameter of the conductors.

Armature reaction

The current that flows in the armature winding also sets up a magnetic field that distorts the flux pattern and tends to reduce the strength of the main magnetic field produced by the field winding. This effect is known as armature *reaction*.

Four-pole d.c. generator

Figure 5.10 shows an end view of the basic construction of a four-pole d.c. generator. Four iron pole pieces are fixed to an iron yoke and the windings on them are wound to ensure that the poles are alternately north, south, north, south in their magnetic polarity. Often the pole pieces are laminated to reduce eddy current losses. The pole pieces are shaped as shown in order to produce a uniform magnetic flux in the air gap between each pole piece and the rotor (armature). The armature core is also often laminated and it is fixed on to the shaft of the machine; the armature coils are placed in slots cut in the surface of the armature (see Fig. 5.9(*b*)).

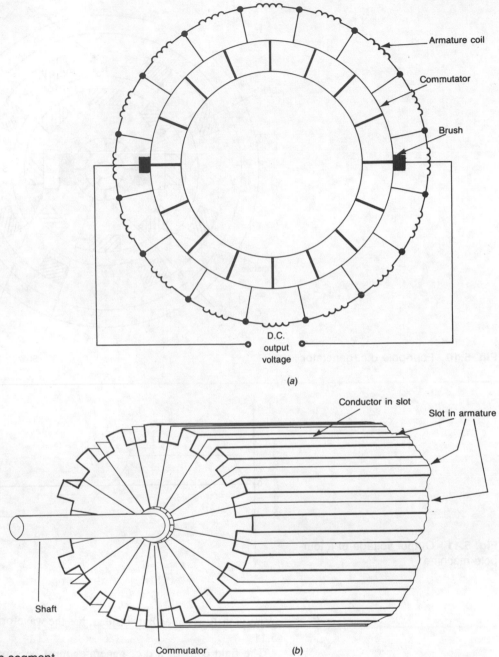

Armature coil

Commutator

Brush

D.C. output voltage

(a)

Conductor in slot

Slot in armature

Shaft

Commutator

(b)

Fig. 5.9 Sixteen-segment commutator: (a) end view; (b) side view

Because of the shaping of the four pole pieces the air gap between the rotor and the stator is of uniform width and this means that the e.m.f. generated in a conductor remains constant for as long as the conductor is alongside a pole piece. The induced e.m.f. then falls rapidly to zero once the conductor moves into the space between two poles. As a result the induced e.m.f. per conductor is not sinusoidal

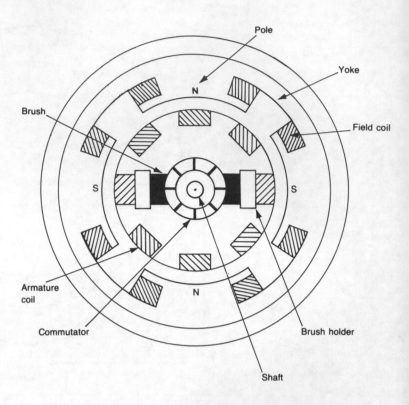

Fig. 5.10 Four-pole d.c. generator

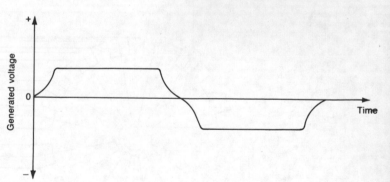

Fig. 5.11 Output voltage of a four-pole machine

as previously assumed but, instead, has the waveform shown by Fig. 5.11.

The field coils of a d.c. generator must be supplied with a d.c. current so that it can provide the necessary magnetic field. It is possible for the field current to be provided by a separate power supply to give a *self-excited d.c. generator*, but it is more convenient for the d.c. generator to provide its own field current. There are two ways in which this can be arranged: (*a*) the field winding can be connected in parallel with the armature — this is a *d.c. shunt generator*; (*b*) the field winding can be connected in series with the armature — this is a *d.c. series generator*. In addition, a combination of the two methods can be used to give a *d.c. compound generator*.

The d.c. shunt generator

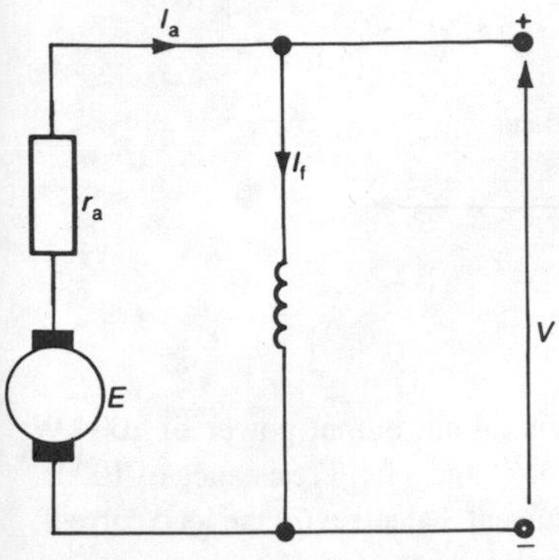

The d.c. shunt generator has its field winding connected in parallel with both its armature and its output terminals as shown by Fig. 5.12. The field winding consists of a large number of turns of small-gauge wire because it does not have to carry the large armature current. When the machine is first started up the generated e.m.f. is very small and entirely due to the residual magnetism in the magnetic circuit. This small e.m.f. causes a small field current to flow and a small magnetic field is set up. This small field, in turn, increases the induced e.m.f. which, in turn, increases the field current and so on. A cumulative effect takes place which results in both the magnetic field and the generated e.m.f. rapidly building up to their correct values for the speed at which the shaft is revolving. The final open-circuit terminal voltage of the generator is determined by the resistance of the field winding and the magnetization curve of the magnetic circuit. The output voltage/field current characteristic is shown in Fig. 5.13.

When the shunt generator is supplying a load there will be an internal voltage drop so that the terminal voltage is less than the generated e.m.f. There are two reasons for this: (*a*) the voltage dropped across the armature resistance r_a, and (*b*) armature reaction. As the current taken by the load increases the terminal voltage falls, slowly at first, but then more rapidly. Eventually a point is reached at which the terminal voltage starts to fall more rapidly than the load current is increased and then the terminal voltage/load current characteristic exhibits a negative slope. This can be seen from Fig. 5.14 which shows how the terminal voltage varies with the load current for a constant shaft speed. The negative characteristic occurs when the fall in the terminal voltage reduces the field current, and hence the magnetic field, to such an extent that the required e.m.f. cannot be generated.

Fig. 5.12 The d.c. shunt generator

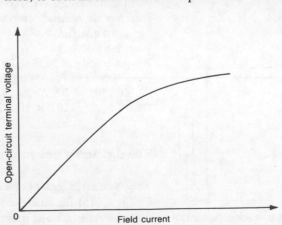

Fig. 5.13 Terminal voltage/field current characteristic of a d.c. shunt generator

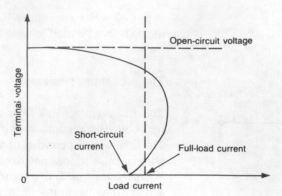

Fig. 5.14 Terminal voltage/load current characteristic of a d.c. shunt generator

Example 5.3

A d.c. shunt generator that supplies a maximum output power of 100 kW at 250 V has an armature resistance of 0.05 Ω and a field resistance of 100 Ω. If the generator is supplying the full load current, calculate (a) the load current, (b) the armature current and (c) the generated e.m.f.

Solution
(a) Load current $I_L = (100 \times 10^3)/250 = 400$ A.   (*Ans.*)
(b) Field current = 250/100 = 2.5 A.
Therefore the armature current = $400 - 2.5 = 397.5$ A.   (*Ans.*)
(c) Terminal voltage $V = E - I_a r_a$. Therefore
$\qquad E = 250 + (397.5 \times 0.05) = 269.88$ V.   (*Ans.*)

Example 5.4

A d.c. shunt generator having an output voltage of 200 V has an armature resistance of 0.4 Ω and a constant field current of 2.5 A. When the speed of the generator is 500 r.p.m. the load is supplied with 2.5 kW power. Calculate the necessary speed for the generator to supply a 4 kW load.

Solution
When the load is 2.5 kW, $200 = E - I_a \times 0.4$. $I_L = 2500/200 = 12.5$ A.
Therefore the armature current $I_a = 12.5 + 2.5 = 15$ A.
Generated e.m.f. $E = 200 + 15 \times 0.4 = 206$ V.
When the load is 4 kW, $I_L = 4000/200 = 20$ A.
The generated e.m.f. $E = 200 + 22.5 \times 0.4 = 209$ V.
The induced e.m.f. is proportional to the speed of the generator and hence
$\qquad 206/500 = 209/N$.
Therefore $N = (209 \times 500)/206 = 507.3$ r.p.m.   (*Ans.*)

The d.c. series generator

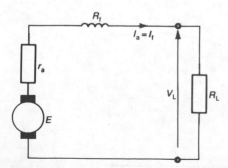

Fig. 5.15 The d.c. series generator

The basic arrangement of a d.c. series-connected generator is shown by Fig. 5.15. Because the field winding is now connected in series with the armature and must carry the armature current it consists of a few turns of heavy-gauge wire. When the output terminals of the generator are open-circuited there can be no field current and so only a small e.m.f., due to the machine's residual magnetism, can be generated. When the generator supplies a load the terminal voltage is initially directly proportional to the armature (=load) current, but at higher currents the output voltage increases less rapidly because the magnetic circuit starts to saturate. The terminal voltage varies with the load current and, since this is usually an undesirable characteristic, the series generator finds little application.

The d.c. compound generator

A d.c. compound generator combines the characteristics of the shunt and the series d.c. generators. The magnetic field is produced by the currents flowing in both high-resistance shunt windings and low-resistance series windings. By the suitable choice of the relative contributions of the two sets of windings a more or less constant terminal voltage can be obtained over a wide range of load currents.

Every year the applications for d.c. generators are diminishing since for increasingly varied applications it is becoming both easier and cheaper to obtain a required d.c. voltage from a rectifier unit that operates from the mains supply.

The d.c. motor

Most small d.c. motors use a permanent magnet to produce the required magnetic field and apply the d.c. voltage via a commutator. Such motors are employed in fixed-speed applications such as tape and cassette recorders, and in variable-speed applications such as electric drills and food mixers. The characteristics of a small d.c. motor are the same as those of the larger types that are to be considered. The direction of rotation can be reversed by reversing the connections of the motor to the voltage supply; this may be done by a simple mechanical switch or an electronic control can be used. A wide variety of electronic control circuits exist for the control of the speed, direction, etc. of a small d.c. motor.

Larger d.c. motors have the required magnetic field produced by the current flowing in the field winding. A d.c. motor is required to rotate when a d.c. voltage is applied across its terminals and this demands the use of a commutator; the action of the commutator is reversed from the action of a d.c. generator's commutator action. The d.c. voltage is applied across two brushes (see Fig. 5.16), and a current flows from the positive terminal, through the commutator segment, through an armature coil, and then back to the negative terminal via the other segment and the other brush. The interaction between the main magnetic field, set up either by the field current or by a permanent magnet, and the field set up by the armature current produces a force that acts upon the coil. Since the coil is free to rotate

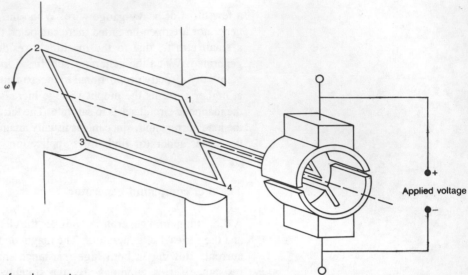

Fig. 5.16 The action of a d.c. motor

it does so and when it has rotated through 180° the other half of the segment is in contact with side 1−2 of the coil and the current flows from the positive terminal, a brush, the other segment, side 4−3 of the coil, and then back to the negative terminal. The forces exerted upon the two sides of the coil are therefore always in the same direction and make the coil rotate in the anticlockwise direction with an angular velocity of ω radians/second.

In a practical d.c. motor, just like a d.c. generator, a greater number than two coils, poles and commutator segments is employed.

When a d.c. voltage is applied to a d.c. motor the motor runs at a particular speed and a back e.m.f. E is generated in the armature winding. The magnitude of this e.m.f. is directly proportional to the rotational speed of the armature and its polarity is such that it opposes the applied voltage. When the motor is on 'no-load' the only mechanical resistance is due to its frictional and windage losses and only a small torque is needed to keep the machine running. Since torque is proportional to current the current taken from the supply is also small. In large motors the friction/windage losses are relatively small and a series-wound motor may have a dangerously high 'no-load' speed. A large series motor should therefore never be run unloaded.

If the mechanical load on the motor is increased the motor will slow down and the induced e.m.f. will fall and the voltage applied across the armature will increase. This increase in voltage will increase the armature resistance current, $[I_a = (V - E)/r_a]$, so that the motor will develop an increased torque to meet the greater load. The speed of the motor will increase until the input electrical power needed to drive the motor is equal to the mechanical output power needed to provide the load torque, i.e. $VI = T\omega$. This means that (a) for a constant electrical input power the output torque is proportional to $1/\omega$, (b) the current taken from the supply by a d.c. motor is a function of the load torque, and (c) all d.c. motors are variable speed machines. The power supply for a large d.c. motor is often supplied by a power electronics circuit that takes a.c. voltage from the mains supply and converts it to the required d.c. voltage.

Example 5.5

A 500 V d.c. motor has an armature resistance of $0.25\,\Omega$ and it runs at 1000 r.p.m. with an armature current of 30 A. When the load on the motor is increased the current increases to 40 A. Calculate the new speed of the motor if the main flux is unaltered.

Solution
$E_1 = 500 - (30 \times 0.25) = 492.5$ V.
$E_2 = 500 - (40 \times 0.25) = 490$ V.
Therefore $492.5/490 = 1000/N$, or
$\qquad N = (490 \times 1000)/492.5 = 995$ r.p.m.   *(Ans.)*

Example 5.6

A 500 V, 10 kW d.c. motor has an armature resistance of $0.08\,\Omega$. The 'no-load' speed is 1000 r.p.m. and the 'no-load' current is 0.6 A. Calculate the 'no-load' speed when the applied voltage is 200 V.

Solution
From equation (5.8) $E = 500 - (0.8 \times 0.6) = 499.52$ V.
From equation (5.4) $k_e = 499.52/1000 = 0.4995$ V/r.p.m.
When $V = 200$ V, $E = 200 - (0.8 \times 0.6) = 199.52$ V. Therefore
$\qquad N = 199.52/0.4995 = 399.4$ r.p.m.   *(Ans.)*

The series motor

A series-connected d.c. motor has its field winding connected in series with its armature shown by Fig. 5.17. This means that the armature current and the field current are the same. The torque developed by a d.c. motor is $T = k_t I_a \Phi$ but now $I_a = I_f$ and Φ is proportional to I_f. Therefore,

$$T = k_t I_a^2 \text{ Nm} \qquad (5.13)$$

i.e. the torque of a d.c. series motor is proportional to the square of the armature current (see Fig. 5.18(*a*)).

At start the speed of a series motor is, of course, zero and so there is no induced e.m.f. The applied voltage will therefore cause a large armature current to flow and this means that the motor has a large starting torque. Further, the induced e.m.f.

$$E = k_e \Phi \omega \quad \text{or} \quad \omega = k_e' E/\Phi = k_e'(V - I_a r_a)/\Phi$$

When the armature current is small, $\omega \propto V/\Phi$ and since the applied voltage V is usually of constant value, $\omega \propto 1/\Phi$. Hence the speed

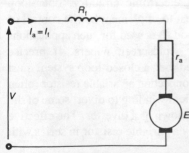

Fig. 5.17 The d.c. series motor

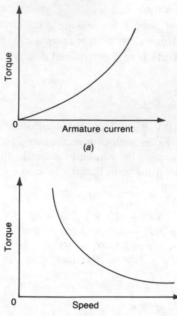

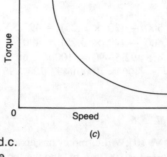

Fig. 5.18 Characteristics of a d.c. series motor: (a) torque/armature current; (b) speed/armature current; (c) torque/speed

of a series-wound d.c. motor is inversely proportional to the armature current. The speed/(armature current) relationship is shown by Fig. 5.18(b). The speed of a series d.c. motor is very high under no-load conditions because then the armature current, and hence the magnetic field, is very small. Large d.c. motors generally have a permanently connected load to avoid any possibility of the speed rising to a dangerously high value. Finally, the torque/speed relationship of a d.c. series motor is shown by Fig. 5.18(c). The characteristics show that the series d.c. motor has a high torque and a low speed at high values of armature current. This type of motor is therefore suitable for use in electric trains, car starter motors, etc, where heavy loads have to be started from rest.

The speed of a d.c. series motor is proportional to $(V - I_a r_a)/\Phi$ and so it can be controlled by varying either the applied voltage, or the flux Φ, or the effective armature resistance. Control of the applied voltage is usually achieved using electronic circuitry; open-loop variable voltage control is the easiest to implement but its control of both speed and torque is not very good. It is used for such applications as electric pumps and motor-car windscreen wipers. If precise selection of motor speeds is required then a closed-loop system must be used. The flux can be varied by connecting a variable resistor (often electronically) in parallel with the field winding to divert some of the field current. This resistor is often known as a diverter. The effective armature resistance can be varied by a variable resistor in series with the armature resistance r_a.

Example 5.7

A 250 V 20 kW series d.c. motor operates at 1200 r.p.m. on full load. Calculate the armature current when the load is reduced until the output torque is 140 N m.

Solution

Full-load power $= I_{FL}V$, and so $I_{FL} = 20\,000/250 = 80$ A.
Also, $20\,000 = $ full-load torque $T_{FL} \times$ angular velocity ω.
Therefore $T_{FL} = (20\,000 \times 60)/(1200 \times 2) = 159.2$ N m.
The torque is proportional to (armature current)^2 and hence
$$159.2/140 = (80/I_a)^2, \text{ or } I_a = 80\sqrt{(140/159.2)} = 75 \text{ A.} \quad (Ans.)$$

Example 5.8

The armature of a d.c. series motor has 500 conductors (250 turns) each of which is 30 cm in length. The diameter of the armature is also 30 cm. When starting, the motor takes a current of 12 A from the supply and this produces a magnetic flux density of 0.6 Wb/m^2. Calculate the starting torque of the motor.

Solution

The tangential force F on each conductor $= BIl = 0.6 \times 12 \times 0.3 = 2.16$ N.
Total force $= 2.16 \times 500 = 1080$ N.
Torque $T = Fr = 1080 \times 0.3/2 = 162$ N m. $\quad (Ans.)$

Universal motor

The *universal motor* is a series-connected d.c. motor that can be operated from either an a.c. or a d.c. voltage supply and it is available at power ratings up to about 500 W. A series d.c. motor has a torque that is proportional to the *square* of the armature current. Hence its torque is always in the same direction regardless of the direction of the current. When such a motor is connected to the 50 Hz mains supply the current reverses its direction every 10 ms and there is a peak in the developed torque 100 times per second, but fluctuations in the speed and torque of the motor are limited by armature reaction. All of the magnetic circuit is laminated and the speed of the machine varies with the load. The operation of the motor is always better when operated from a d.c. supply but it is perfectly adequate for such applications as vacuum cleaners, food mixers and other household appliances. A universal motor runs at a speed somewhere between 8000 and 12 000 r.p.m. and its speed can be controlled by a simple triac circuit which varies the applied voltage. Variable speed universal motors are employed in electric drills and fans.

The d.c. shunt motor

Figure 5.19 shows how the field winding of a shunt motor is connected

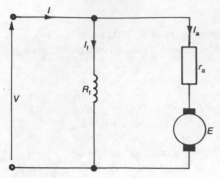

Fig. 5.19 The d.c. shunt motor

in parallel with the armature. Since the d.c. supply voltage is applied directly across the field winding the field current is constant. This does not mean, however, that the magnetic field is constant since armature reaction will reduce the field and this increases the speed of the motor. Also, as the armature current increases the induced e.m.f. E will fall because of the increased $I_a r_a$ voltage drop, and this tends to decrease the speed of the motor. The two effects tend to cancel out and so the speed of a shunt motor is almost constant over a wide range of load torques and it is hence suitable for such applications as electric pumps, compressors and lifts. The speed/armature current characteristic is shown in Fig. 5.20(a).

The speed of a d.c. shunt motor is proportional to $(V - I_a r_a)/\Phi$ and hence it can be controlled by varying either the applied voltage V, or the flux Φ, or the effective armature resistance r_a. The applied voltage can be varied using various electronic control circuits that are beyond the scope of this book. The flux can be varied by connecting a variable resistor in series with the field winding (known as a field regulator) which will control the value of the field current and hence the flux. The effective value of the armature resistance can be increased by inserting a variable resistor in series with the armature.

The torque T of the shunt motor is equal to $T = k_t \Phi I_a$ and since Φ is constant the torque increases, more or less linearly, with the armature current until armature reaction causes the main magnetic field to be reduced (see Fig. 5.20(b)). The torque $T = EI_a/(2\pi N)$ and so the torque/speed characteristic is as shown by Fig. 5.20(c).

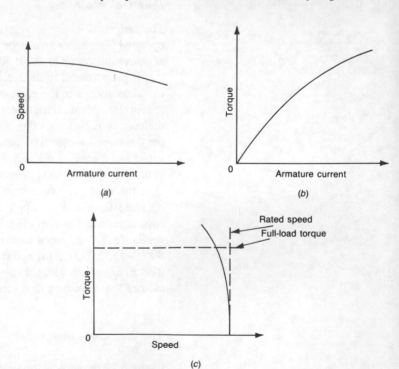

Fig. 5.20 Characteristics of a d.c. shunt motor: (a) speed/armature current; (b) torque/armature current; (c) torque/speed

Example 5.9

A 150 V d.c. shunt motor has an armature resistance of 0.4 Ω. The no-load and full-load armature currents are 10 and 90 A respectively, and the no-load speed is 300 r.p.m. Calculate the speed of the motor at full load.

Solution

On no-load, $E = 150 - (10 \times 0.4) = 146$ V.

On full-load, $E = 150 - (90 \times 0.4) = 114$ V.

Now $E = kN$, so $146 = 300k$ and $k = 146/300$. Therefore

$\quad N = 114/k = (114 \times 300)/146 = 234.3$ r.p.m.   (*Ans.*)

Example 5.10

A 250 V d.c. shunt motor has an armature resistance of 0.4 Ω and a field resistance of 200 Ω and when the load torque is 100 N m takes a current of 20 A from the supply. Calculate the speed of the motor. Calculate the new speed if the load torque is increased to 120 N m.

Solution

Field current $I_f = 250/200 = 1.25$ A.

The armature current $I_a = 20 - 1.25 = 18.75$ A.

Hence, $E = 250 - (18.75 \times 0.4) = 242.5$ V.

Torque $= EI_a/2\pi N = (242.5 \times 18.75)/(2\pi \times 100)$

$\qquad\qquad\quad = 7.24$ r.p.s $= 434.2$ r.p.m.   (*Ans.*)

When the load torque is increased to 120 N m the armature current is increased in proportion to $18.75 \times (120/100) = 22.5$ A. Then

$\quad E = 250 - (22.5 \times 0.4) = 241$ V.

Therefore the new speed $= (241 \times 22.5)/(2\pi \times 120)$

$\qquad\qquad\qquad\qquad\quad = 7.19$ r.p.s. $= 431.5$ r.p.m.   (*Ans.*)

Starting a d.c. motor

When a d.c. motor is stationary the induced e.m.f. E will be zero and so when a voltage V is applied to the motor's terminals the initial current that flows will be equal to V/r_a. Since r_a is a small value of resistance this initial current will be much larger than the running current; this may not be a problem with small motors, but it most certainly will be for all large motors where damage will be caused unless the initial current is limited in some way. The necessary limitation of the initial armature current is obtained by connecting external resistance R_s in series with the motor. The basic circuit of a *starter* is shown in Fig. 5.21; at any instant the armature current $I_a = (V - E)/(R_s + r_a)$. If, for example, $V = 400$ V and $r_a = 0.4 \Omega$ the initial current without a starter resistance would be 100 A. If the starting current is to be limited to 50 A the initial total resistance $R_s + r_a$ must be $400/50 = 8 \Omega$ and so the total starter resistance should be 7.6 Ω. This could be divided into four equal resistances of 1.9 Ω. As the motor increases speed the induced e.m.f. E becomes larger

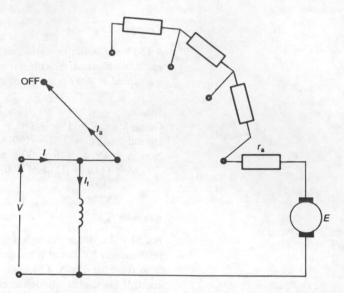

Fig. 5.21 Series motor starter

and R_s can be reduced until, when the motor has reached its final speed, its resistance is zero.

Stopping (braking) a d.c. motor

To stop, or brake, a d.c. motor the applied voltage must be switched off. The time taken for the motor to come to rest is determined by its inertia and by the friction and windage losses. Often, this time will be excessively long and if faster braking is wanted further steps must be taken:

(a) *Dynamic or resistive braking*: When the motor is disconnected from the voltage supply it is immediately connected to a resistor R in which the inertial energy can be rapidly dissipated (see Fig. 5.22(a)).

(b) *Reverse current*: This is shown in Fig. 5.22(b); the connections of the motor to the voltage supply are switched over to reverse the direction of the armature current. This produces a reverse torque that rapidly slows the motor down.

(c) *Regenerative*: With this method of braking, shown in Fig. 5.22(c), the field is adjusted to make the induced e.m.f. E greater than the applied voltage V. The machine then acts as a generator returning energy from the mechanical system to the supply and rapidly slowing the machine down.

The induction motor

Large induction motors operate from a three-phase voltage supply and their action depends upon the production of a rotating magnetic field. The current-carrying conductors are mounted on the stator and this means that there are no external connections to the rotor and hence

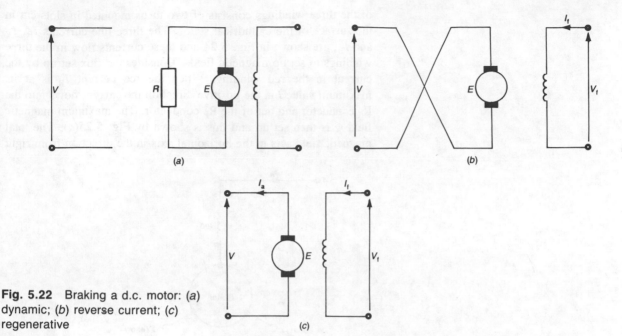

Fig. 5.22 Braking a d.c. motor: (a) dynamic; (b) reverse current; (c) regenerative

neither slip-rings nor a commutator are required. This results in a machine that is relatively cheap, more robust and reliable than commutator-type motors and so needs little maintenance, and is self-starting. The three-phase induction motor is commonly employed in industry to operate a wide variety of industrial drives, large pumps, compressors, conveyor belts, etc. The speed of an induction motor is determined by the frequency of the power supply. If a variable-frequency d.c./a.c. converter is employed the induction motor can be used in controlled-speed applications.

Small induction motors are employed in various domestic appliances, such as washing machines and electric fans, and these must be able to operate from a single-phase voltage supply.

Rotating magnetic field

A rotating magnetic field can be produced by applying a three-phase voltage supply to three windings that are mutually spaced 120° apart. The three windings, red, yellow and blue, may be either star-connected, as shown by Fig. 5.23(a), or delta-connected. The positive direction of currents in the coils is taken to be from a supply line to the stator, or from the start (S) end of a winding to the finish (F) end. Each of the currents flowing in a winding sets up a magnetic field around that winding and the total flux created at any instant in time is equal to the phasor sum of the three individual magnetic fluxes.

Figure 5.23(b) shows an end view of a simple stator in which each of the three windings consists of two turns mounted in slots cut in the surface of the cylindrical stator. The three line currents I_R, I_Y and I_B, are shown in Fig. 5.24 and these currents flow in the three windings to set up magnetic fields. Consider the flux set up by the current in the red winding. When the red current I_R is at its maximum value I in the positive direction the current flows into the R_S conductor and out of the R_F conductor. The maximum magnetic field Φ is then set up and this is shown by Fig. 5.25(a). The total magnetic field acts in the horizontal axis in the direction from right

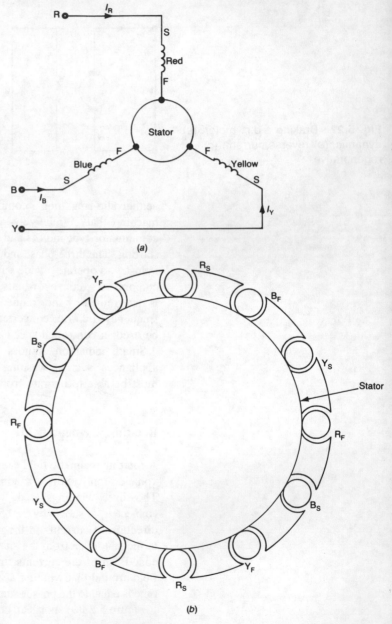

Fig. 5.23 Induction motor: (a) three-phase windings; (b) end view of stator

to left. One-twelfth of a cycle later, or 30°, the red current is still positive but its amplitude has fallen to 0.866I. The resultant magnetic flux is now 0.866Φ but it still acts along the same axis in the same direction as shown by Fig. 5.25(*b*). After another 30° phase lag the amplitude of the current has fallen to 0.5I and so the resultant magnetic flux is now 0.5Φ but is still in the same, positive, direction (Fig. 5.25(*c*)). Another 30° later brings the amplitude of the red current to zero and then there is no magnetic flux produced (Fig. 5.25(*d*)). Thereafter the current starts to increase in the negative direction; the flux set up is now along the same axis but in the opposite direction as shown by Figs 5.25(*e*)–(*h*). If, therefore, a sinusoidal current is passed through one of the windings an alternating magnetic flux will be produced that will always act along the same axis.

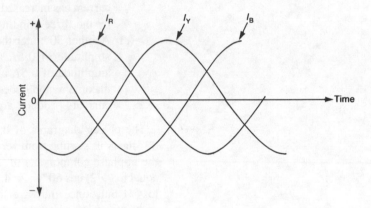

Fig. 5.24   Line currents

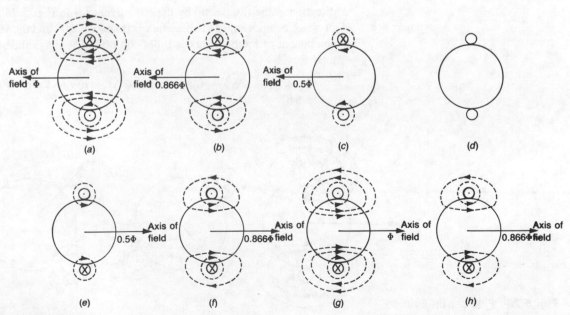

Fig. 5.25   Generation of a single-phase magnetic field

When all three of the windings are considered the magnetic flux produced by each of them can be determined in similar fashion. The total combined magnetic flux will be of constant amplitude and it will rotate around the stator at a constant velocity.

Referring to the current waveform given in Fig. 5.24:

(a) At the instant when the red current I_R is at its maximum positive value I, the yellow and blue currents, I_Y and I_B, are each negative and of amplitude $0.5I$. The magnetic fields produced by each of the currents acting on its own are shown by Fig. 5.26.

(b) 30° later the red current is still flowing in the positive direction but its amplitude has fallen to $0.866I$. The yellow current is now zero and the amplitude of the negative blue current has increased to $0.866I$. The magnetic fields produced by the three windings are now as shown by Fig. 5.27(a).

(c) Another 30° later the red current is still positive and has an amplitude of $0.5I$, the yellow current is positive with an amplitude of $0.5I$, and the blue current is negative with its maximum amplitude of I. The magnetic fields that these three currents produce are shown by Fig. 5.27(b).

The phasor diagrams of the magnetic fields at three consecutive 30° intervals starting from zero are given by Fig. 5.28. In Fig. 5.28(a) the in-phase components of the yellow and the blue fluxes are each equal to $(\Phi/2) \cos 60° = \Phi/4$. The vertical components are also equal to $\Phi/4$ but, since they are in opposite directions, they cancel one another out. Hence the total flux Φ_T is equal to 1.5Φ in the same direction as the flux set up by the red current I_R. In Fig. 5.28(b) the in-phase component of the blue flux is $0.866 \cos 60°$ and the vertical component of the blue flux is $0.866 \sin 60°$. The magnitude of the total flux Φ_T is

$$\Phi_T = \sqrt{[(0.866\Phi + 0.866\Phi \cos 60°)^2 + (0.866\Phi \sin 60°)^2]} = 1.5\Phi$$

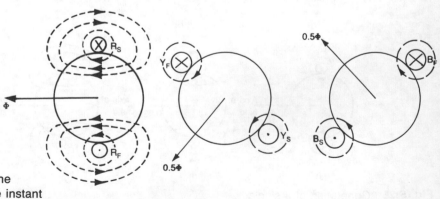

Fig. 5.26 Flux set up by the currents I_R, I_Y and I_B at the instant I_R is at its positive maximum value

(a)

(b)

(c)

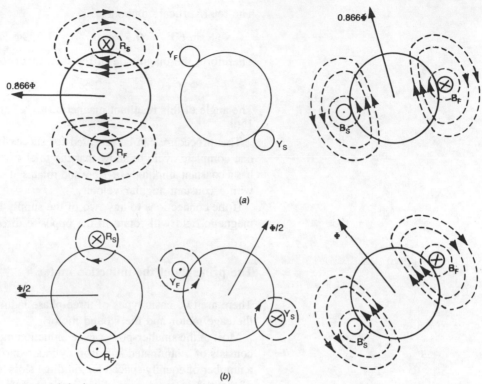

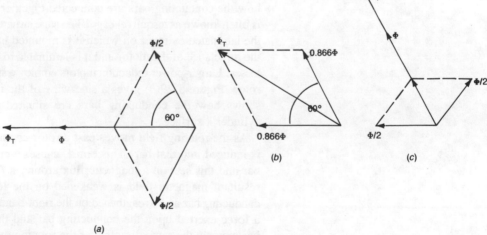

Fig. 5.27 Flux set up when (a) I_R is 0.866 times its maximum value and (b) I_R is 0.5 times its maximum value

Fig. 5.28 Phasor diagram of the magnetic field set up by I_R, I_Y and I_B

The angle of the total flux is

$$\tan^{-1}[(0.866 \sin 60°)/(0.866 + 0.866 \cos 60°)] = 150°$$

In Fig. 5.28(c) the total horizontal component of the flux is

$$\Phi(-0.5 - 1 \cos 60° + 0.5 \cos 60°) = -0.75\Phi$$

The total vertical component is

$$\Phi(1 \sin 60° + 0.5 \sin 60°) = 1.299\Phi$$

Therefore, the magnitude of the total magnetic flux is

$$\Phi_T = \Phi \sqrt{(-0.75^2 + 1.299^2)} = 1.5\Phi.$$

The angle of this resultant magnetic flux is $\tan^{-1}(1.299/-0.75) = 120°$.

This procedure can be repeated for successive 30° intervals over one complete cycle to show that the total combined magnetic field has a constant amplitude of 1.5Φ and rotates in the clockwise direction with a constant angular velocity.

If the connections to any two of the supply lines are reversed the magnetic field will rotate in the opposite direction.

The principle of the induction motor

There are two basic types of three-phase induction motor known as the cage motor and the wound motor.

Most of the smaller-power cage induction motors have a rotor that consists of a laminated soft iron cylinder into the surface of which a number of equally spaced longitudinal slots have been cut. In each of the slots a copper, or aluminium, bar is placed and the ends of each bar are short-circuited together by two conducting end-rings. The basic construction is shown by Fig. 5.29. Figure 5.29(a) shows how the conducting bars are connected together to form a cage that is often known as a squirrel cage. The cage structure is the rotor itself, the laminated cylinder on which it is mounted is provided to reduce the reluctance of the motor and it is laminated to reduce eddy current losses. Larger-power induction motors employ a star-connected wound rotor. Figure 5.29(b) gives a side view of the induction motor and shows how the conducting bars are situated inside slots on the cylinder's surface.

As the rotating field moves past a rotor conducting bar an e.m.f. is induced into that bar. This e.m.f. causes a current to flow in the bar and this sets up a magnetic flux around it (see Fig. 5.30). The resultant magnetic field is weakened on the left-hand side of the conducting bar and strengthened on the right-hand side. There is hence a force exerted upon the conducting bar and the torque makes the bar move in the same direction as the rotating magnetic field. As the rotor picks up speed the relative velocity between the rotor and the rotating magnetic field is reduced and hence the e.m.f. induced into the conducting bar is also reduced. In turn, the current flowing in the bar becomes smaller, and since torque is proportional to current, the torque developed by the motor also falls. This means that the greater the speed of the motor the smaller will be the torque that the motor is able to develop.

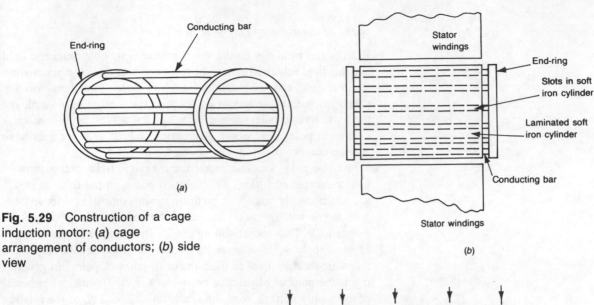

Fig. 5.29 Construction of a cage induction motor: (a) cage arrangement of conductors; (b) side view

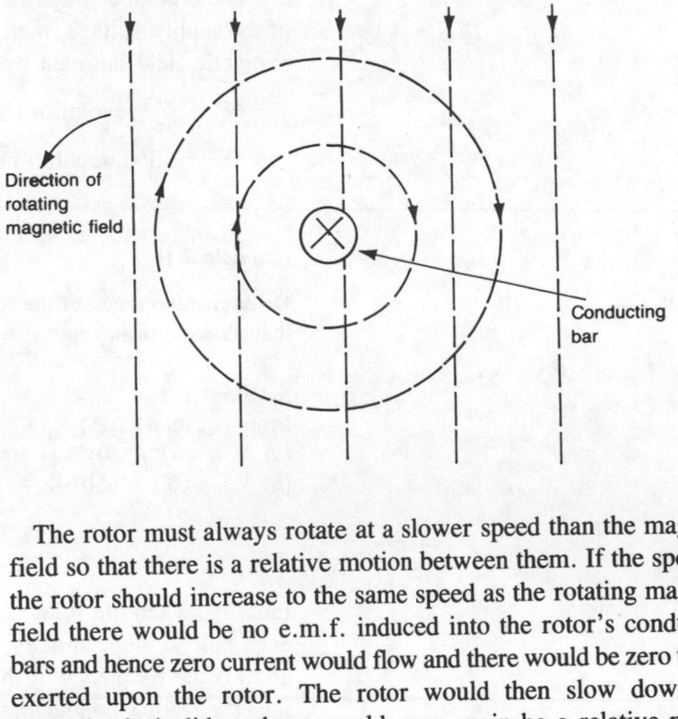

Fig. 5.30 How a force is exerted upon the cage rotor

The rotor must always rotate at a slower speed than the magnetic field so that there is a relative motion between them. If the speed of the rotor should increase to the same speed as the rotating magnetic field there would be no e.m.f. induced into the rotor's conducting bars and hence zero current would flow and there would be zero torque exerted upon the rotor. The rotor would then slow down and immediately it did so there would once again be a relative motion between the magnetic field and the rotor and an e.m.f. would again be induced into the conducting bars. The difference in speed between the speed of the magnetic field, which is known as the synchronous speed, and the speed of the rotor, is known as the *slip*.

The wound induction motor has a rotor with a three-phase winding similar to that for the stator and which is connected to the slip-rings mounted on the rotor shaft. This type of induction motor can give both a high starting torque and good speed regulation.

Speed of the rotating magnetic field

Each of the windings on the stator produces its own magnetic field and this field acts right across the rotor and appears to originate from a north pole and terminate on a south pole. The three-winding arrangement thus produces two poles. If six windings are employed there will be four poles, nine windings give six poles, twelve windings give eight poles, and so on. A practical induction motor may have a large number of windings and hence of poles.

In a two-pole induction motor the magnetic field rotates through one complete revolution, i.e. past two poles, in the time taken for the power supply voltage to go through one complete cycle. A four-pole motor will need two cycles of the supply voltage before the magnetic field has rotated through 360°. In fact, the number of cycles of the supply voltage needed for one complete revolution of the magnetic field is equal to the number of *pairs* of poles. In general, if p is the number of pairs of poles and f is the frequency, in hertz, of the supply voltage, then the synchronous speed N_s of the rotating magnetic field is given by

$$N_s = f/p \text{ revolutions/second} \tag{5.14}$$

$$= 60f/p \text{ revolutions/minute} \tag{5.15}$$

Example 5.11

Determine the speed of the rotating magnetic field in a three-phase 50 Hz induction motor having (*a*) 6 poles and (*b*) 16 poles.

Solution
From equation (5.15),
(*a*) $N_s = (60 \times 50)/3 = 1000$ r.p.m.　　(*Ans.*)
(*b*) $N_s = (60 \times 50)/8 = 375$ r.p.m.　　(*Ans.*)

Slip

Initially, when the rotor is stationary the magnetic field moves past each pole at a speed of pf, where p is the number of pairs of poles and f is the frequency of the supply. At this time the e.m.f. induced into the rotor's conductors is at the same frequency as the voltage supply. As the rotor begins to rotate in the same direction as the rotating magnetic field the frequency of the field relative to the rotor falls. The frequency of the induced e.m.f. is then equal to $f_s - f_r$, where f_s is the frequency of the voltage supply and f_r is the frequency of the rotor's rotation. Further, the rotor current is also reduced and this reduces the torque applied to the rotor.

The slip is the difference between the synchronous speed and the speed of the rotor.

$$\text{Slip} = (N_s - N_r) \tag{5.16}$$

The percentage slip is

$$\text{slip/(synchronous speed)} \times 100\% = [(N_s - N_r)/N_s] \times 100\%$$
$$(5.13)$$

Example 5.12

A three-phase 50 Hz induction motor has four poles and 2% slip. Calculate (*a*) the synchronous speed, (*b*) the speed of the rotor and (*c*) the frequency of the e.m.f. induced into the rotor.

Solution

(*a*) $f_s = f/p = 50/2 = 25$ Hz.   (*Ans.*)
(*b*) $0.02 = (50 - f_r)/50$, or $f_r = 49$ r.p.s. $= 2940$ r.p.m.   (*Ans.*)
(*c*) The rotating field cuts the rotor conductors at a frequency of
    $50 - 49 = 1$ Hz.   (*Ans.*)

Example 5.13

A four-pole 50 Hz three-phase induction motor runs at 1450 r.p.m. Calculate its slip.

Solution

$N_s = (60 \times 50)/2 = 1500$ r.p.m.
Slip $= (1500 - 1450)/1500 = 0.033 = 3.33\%$.   (*Ans.*)

Typical values for slip are between about 3.5 and 6% for small motors and about 1.5−2.5% for large motors. On no load the torque exerted on the rotor only has to overcome the internal friction and windage losses and the no-load slip may be as small as 1%.

Slip/load characteristics

When there is zero mechanical load on an induction motor the internal losses, due to friction and windage, are small and the speed of the rotor is very nearly equal to the synchronous speed. As the mechanical load on the motor is increased the speed of the rotor decreases and both the frequency and the magnitude of the e.m.f. induced into the rotor increase and so the current in, and torque exerted on, the rotor both increase. This means that the slip of an induction motor will increase with increase in the mechanical load on the motor. Figure 5.31 shows the shape of the torque/speed characteristic of an induction motor. When the machine is stationary the torque applied to the rotor is small but it increases as the motor increases speed until the point is reached at which the maximum torque is developed. Beyond this point the e.m.f. induced into the rotor decreases and the torque rapidly falls, becoming zero at the synchronous speed. Normally, an induction motor operates over the nearly constant speed portion of the characteristic as indicated in Fig. 5.31.

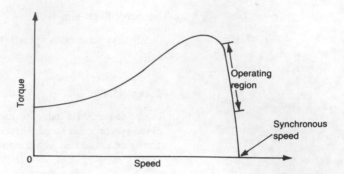

Fig. 5.31 Torque/speed characteristic of an induction motor

Small motors

The electric motors used in the home and in offices and shops are of small physical size and small electrical output power. Both a.c. and d.c. motors are employed; the former operate directly from the 50 Hz public mains supply, the latter from a suitable d.c. power supply such as a battery or a rectifier unit. Another type of motor, known as a universal motor, is able to operate from both a.c. and d.c. power supplies with a speed that varies with the load in the same way as a d.c. series motor. Although the universal motor tends to work somewhat better from a d.c. supply than an a.c. supply, its performance is perfectly adequate for such domestic applications as vacuum cleaners, food mixers and hair driers.

Since the domestic power supply is 230 V single-phase the induction motors that are used in the home, for such purposes as refrigerators and washing machines, are all single-phase types. It is not possible to start an induction motor from a single-phase voltage supply, since this cannot produce the necessary rotating magnetic field, although it will continue to run once it has been started. An induction motor will, however, start from a two-phase supply, which can produce a rotating field, and so some method is required to provide a two-phase supply during the start-up period. There are two basic ways of doing this, known, respectively, as the *split-phase* and *shaded-pole* motors.

One version of a split-phase induction motor is the *capacitor-start* motor, shown in Fig. 5.32, which is used for power outputs of up to about 500 W. This has two windings on its stator; the start winding is connected to the supply via a capacitor and is only in circuit while a centrifugal switch is closed. Because of the capacitor the current that flows in the start winding is out of phase (leading) with the current that flows in the running winding. The magnetic fields set up by the windings are out of phase and combine to produce the rotating field necessary for the motor to start. The motor starts as a two-phase induction motor. When the squirrel cage rotor has reached its operating speed the centrifugal switch operates and cuts both the capacitor and the start winding out of the circuit and now the motor runs as a single-phase machine. This type of motor has a performance that is not far short of that of a three-phase motor and it is used, for example, in refrigerators and freezers, and any other applications where long-term operation is wanted.

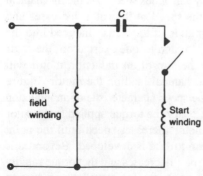

Fig. 5.32 Capacitor-start single-phase induction motor

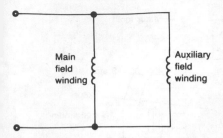

Fig. 5.33 Split-phase single-phase induction motor

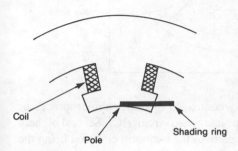

Fig. 5.34 Shaded-pole single-phase induction motor

The stepper motor

In many modern motors there is no centrifugal switch and the two windings are in use all the time. Then the main winding uses thick-diameter wire so that it is of low resistance, while the auxiliary winding has only a few turns of fine wire and is hence of higher resistance (see Fig. 5.33). The difference between the impedances of the two windings is all that is necessary to produce the required phase difference between the two currents for a two-phase rotating field to be set up. This arrangement gives a higher starting torque than does the capacitor-start motor.

A shaded-pole induction motor is only used for low output power applications. The stator has a single-phase winding that is usually wound on salient poles that have a copper shading ring fitted around one-half of a pole, as shown by Fig. 5.34. The effect is to produce a phase lag in a part of the motor's magnetic circuit. When a current flows in the winding the flux set up induces an e.m.f. into the shading ring and this causes a current to flow in it. In turn this current sets up its own magnetic field and this produces the required two-phase magnetic field. The rotating field produced is not a very good one and this results in the motor having a rather poor performance. Also its efficiency is low because there is a continuous power loss in the shading ring, but because of the motor's small size this is not important. This type of induction motor is relatively simple and hence cheap to manufacture but it can only be used for applications where the starting torque is light and they are used, for example, in hair driers.

Stepper motors are used in robotics, various control systems, computer printers, electronic typewriters, disc drives and many other microprocessor/computer-controlled equipments. The stepper motor gives precise control of the rotation, position, speed and direction of the controlled item. It does not continuously rotate, like other motors, but moves in a series of small steps. Each time the motor is supplied with a current pulse it moves through a fixed angle.

There are two main types of stepper motor: (*a*) the variable reluctance type and (*b*) the hybrid type. The former uses the current flowing in windings to produce the magnetic field, and the latter uses one or more permanent magnets attached to the rotor and four separate windings around the stator. For either type of stepper motor when d.c. current pulses are passed through the windings in a particular way the rotor is turned through a particular angle. Some stepper motors have an angle of rotation per step of 7.5° or 15° while more expensive motors will have smaller steps of, say, 1.8°. By the application of a suitable sequence of pulses the output shaft of the motor can be turned through a precise number of steps, either backwards or forwards. Each step is completed in a few milliseconds. The output shaft can also be made to rotate continuously at any wanted speed (within limits, of course) in either direction.

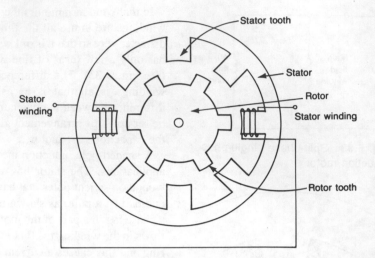

Fig. 5.35 A variable reluctance stepper motor

A stepper motor is usually controlled by either a microprocessor or a specialized stepper motor integrated circuit (IC). Several of these ICs are on the market, with perhaps the two most common being the SAA 1024 and the SAA 1027.

The basic construction of a variable-reluctance stepper motor is shown by Fig. 5.35. It consists of an eight-tooth stator and a soft iron rotor having six teeth mounted inside of the stator. All the rotor teeth and all the stator teeth are of the same width and there is a small air gap between them. The winding on each stator tooth is wound in the opposite sense to the winding on the opposite stator tooth and opposite pairs are connected in series. Two such opposite teeth windings form one *phase* with one winding acting as a north pole and the other as a south pole. This means that there are 8/2 = 4 phases in this motor. When a current pulse is passed through any phase winding it sets up a magnetic field that goes from the north tooth across the low-reluctance rotor to the south tooth. Lines of magnetic flux always try to shorten their length and so a force is exerted upon the rotor that causes it to rotate until a pair of rotor teeth are lined up with a pair of stator teeth.

If the two windings shown in Fig. 5.35, say phase A, have a current passed through them, the magnetic field set up passes from the left-hand stator tooth to the right-hand stator tooth and in so doing exerts a force upon the rotor. This force causes the rotor to move until the flux lines are straight and hence of minimum length. If then, the current is removed from phase A and applied to phase B instead the rotor will be forced to move through 15° when two of its teeth will line up with two of the stator teeth. (One complete revolution = 360°/n = 360/8 = 45°. There are three pairs of rotor teeth and so a step from one tooth to the next = 45°/3 = 15°.) If, now, the next phase is energized the rotor will step another 15° and so on. The rotor can easily be stepped in the opposite direction by reversing the sequence

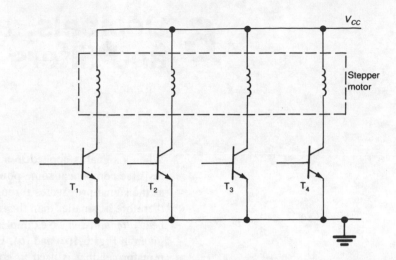

Fig. 5.36 Supplying a stepper motor

Table 5.1

Step	Winding A	Winding B	Winding C	Winding D
1	ON	OFF	ON	OFF
2	OFF	ON	ON	OFF
3	OFF	ON	OFF	ON
4	ON	OFF	OFF	ON

in which the current is passed through the stator windings. Continuous rotation is achieved by continuously repeating the correct pulsing sequence. One way in which the stator windings can be supplied with current pulses is shown by Fig. 5.36. Any of the transistors can be turned ON by a positive base voltage and this transistor then supplies a current to the winding that is connected in its collector circuit. Table 5.1 shows the sequence in which the windings should have a current passed through them for the motor to rotate.

The more commonly employed hybrid stepper motor works in a similar manner but its rotor incorporates a permanent magnet.

6 Decibels, attenuation and filters

When a signal is applied to a circuit, or a network, that contains a resistive component some power will be dissipated in the resistance as the signal propagates through the circuit. The output power will therefore be smaller than the input power and so the input signal has been *attenuated*. Two examples of networks that attenuate signals are given in Figs 6.1(a) and (b). Figure 6.1(a) represents an iron-cored transformer that is used to couple a resistive load R_L to a voltage source of e.m.f. E_s and internal resistance R_s. For maximum power transfer to take place the turns ratio n of the transformer should be $n = \sqrt{(R_s/R_L)}$ (p. 59). Both the primary and the secondary windings of the transformer possess inevitable self-resistance and so the currents I_p and I_s that flow in the windings will produce I^2R power losses. In addition, there are eddy current and hysteresis losses in the

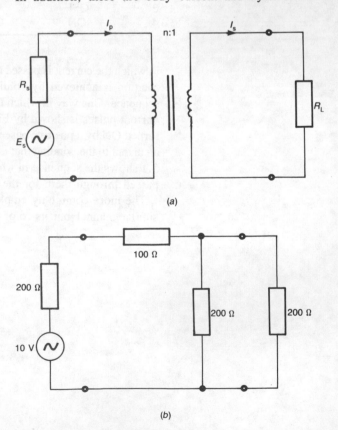

(a)

(b)

Fig 6.1 Attenuation in (a) a transformer circuit and (b) an L-network

transformer's core. This means, of course, that the power output of a transformer will always be less than the input power. The input resistance of the circuit given in Fig. 6.1(b) is $100 + 200/2 = 200\,\Omega$ and hence a voltage of 5 V will appear across its input terminals. The input power to the network is equal to $25/200 = 125\,mW$. The input current is $10/400 = 25\,mA$ and the current that flows into the load resistance is $25/2 = 12.5\,mA$. Hence the power dissipated in the load is $(12.5 \times 10^{-3})^2 = 31.25\,mW$. Clearly, the output power is smaller than the input power and so a power loss has occurred.

In heavy-current engineering where the actual power dissipated at a point is of prime importance, it is usual to refer to the efficiency $100\,P_{out}/P_{in}$ of a circuit. In light-current engineering, however, the *ratio* of the output power to the input power is usually of greater importance and so a unit known as the *decibel* is employed to express power losses (or gains).

The loss introduced by a network is known as the *attenuation* of that network and it is usually quoted in decibels. A network that has been designed to introduce a given amount of attenuation is known as an *attenuator*. The symbol for an attenuator is given in Fig 6.2, the number of decibels loss provided by the network is marked inside the symbol. If, of course, the output power from a circuit is larger than the input power a *gain* has occurred and then the circuit must be some kind of amplifier.

In communication systems in particular, and also in some electronic circuits, there is often a need for a circuit that will permit a band of frequencies to pass while at the same time suppressing all other frequencies that lie outside of that bandwidth. A circuit that performs this function is known as a *filter*. A filter has an attenuation that varies with frequency in the manner demanded by its particular application. There are four basic kinds of filter, known respectively as the low-pass filter, the high-pass filter, the band-pass filter, and the band-stop filter. The symbols for these four filters are given in Fig. 6.3.

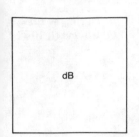

Fig. 6.2 Symbol for an attenuator

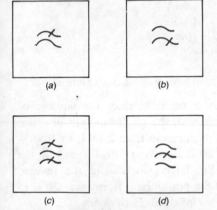

Fig. 6.3 Symbols for filters: (a) low-pass; (b) high-pass; (c) band-pass; (d) band-stop

The decibel

The decibel (dB) is defined as ten times the logarithm to base 10 of the ratio of two *powers*. Thus, if the input power to a circuit is P_{in} watts and the output power is P_{out} watts then, if $P_{out} > P_{in}$, the *gain* G of the circuit is

$$G = 10 \log_{10}(P_{out}/P_{in})\ dB. \tag{6.1}$$

If $P_{in} > P_{out}$ the *loss* L of the circuit is

$$L = 10 \log_{10}(P_{in}/P_{out})\ dB. \tag{6.2}$$

Example 6.1

When an input power of 10 mW is applied to a power amplifier the output power is 5 W. Calculate the gain of the amplifier.

Solution

Gain $= 10 \log_{10} [5/(10 \times 10^{-3})] = 27$ dB.   (*Ans.*)

Example 6.2

Convert each of the following power ratios into decibels: $P_{out}/P_{in} = $ (*a*) 2, (*b*) 4, (*c*) 8, (*d*) 1000, (*e*) 2000, (*f*) 10^6, (*g*) 4×10^6, (*h*) 0.1 and (*i*) 10^{-3}.

Solution

(*a*) $G = 10 \log_{10} 2 = 3.01$ dB $\simeq 3$ dB.   (*Ans.*)
(*b*) $G = 10 \log_{10} 4 = 6.02$ dB $\simeq 6$ dB.   Ans)
(*c*) $G = 10 \log_{10} 8 = 9.03$ dB $\simeq 9$ dB.   (*Ans.*)
(*d*) $G = 10 \log_{10} 1000 = 30$ dB.   (*Ans.*)
(*e*) $G = 10 \log_{10} 2000 = 33$ dB.   (*Ans.*)
(*f*) $G = 10 \log_{10} 10^6 = 60$ dB.   (*Ans.*)
(*g*) $G = 10 \log_{10} (4 \times 10^6) = 66$ dB.   (*Ans.*)
(*h*) $G = 10 \log_{10} 0.1 = -10$ dB.   (*Ans.*)
or $L = 10 \log_{10} 1/0.1 = 10$ dB.   (*Ans.*)
(*i*) $G = 10 \log_{10} 10^{-3} = -30$ dB.   (*Ans.*)
or $L = 10 \log_{10} 10^3 = 30$ dB.   (*Ans.*)

A number of useful points should be noted from example 6.2:

(*a*) A doubling of the power ratio increases the number of decibels by 3 dB regardless of the size of the power ratio. Thus, increases in the power ratio from 2 to 4, or from 4 to 8, or from 1000 to 2000 all give a 3 dB increase in the number of decibels. Similarly, working in the reverse direction, a halving of the power ratio from, say, 2000 to 1000 reduces the number of decibels by 3 dB.

(*b*) A fourfold increase in the power ratio, e.g. from 2 to 8, gives a 6 dB increase in the number of decibels.

(*c*) A power ratio less than unity gives a negative decibel value and this must be interpreted as a loss, i.e. the output power is less than the input power. A loss can always be calculated as $10 \log_{10}$(input power)/(output power).

Example 6.3

A 20 mW signal is applied to the input terminals of a 6 dB attenuator. Calculate the output power.

Solution

$6 = 10 \log_{10} (20/P_{out})$
$10^{0.6} = 3.98 = 20/P_{out}$, or
$\quad P_{out} = 20/3.98 = 5.025$ mW.   (*Ans.*)
Alternatively, $-6 = 10 \log_{10} (P_{out}/20)$
$10^{-0.6} = 0.2512 = P_{out}/20$, or
$\quad P_{out} = 20 \times 0.2512 = 5.024$ mW.   (*Ans.*)

The small difference between the two answers has occurred because some of the figures have been rounded off. In practice, 6 dB is usually taken as being equal to a power ratio of 4, or 0.25, and then P_{out} = 20/4 = 5 mW, or P_{out} = 20 × 0.25 = 5 mW.

Cascaded stages

When a number of stages are connected in cascade the overall gain, or loss, of the system can be determined by merely adding the decibel values of the individual gains, or losses, of each stage.

Example 6.4

Calculate the overall gain, or loss, of the system shown in Fig. 6.4. If the input power is 5 mW calculate the output power.

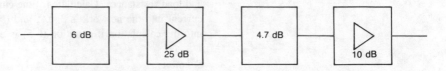

Fig. 6.4

Solution
Overall gain = −6 + 25 − 4.7 + 10 = 24.3 dB.   (*Ans.*)
$24.3 = 10 \log_{10}[P_{out}/(5 \times 10^{-3})]$
$10^{2.43} = 269.2 = P_{out}/(5 \times 10^{-3})$, or
$\quad P_{out} = 269.2 \times 5 \times 10^{-3} = 1.346$ W.   (*Ans.*)

Voltage and current ratios

The decibel is defined as being equal to $10 \log_{10}(P_{out}/P_{in})$ and, since the power dissipated in a resistance may be written as either I^2R or V^2/R, the gain, or the loss, of a circuit may be written as either

$$G \text{ (or } L) = 10 \log_{10}[(I_{out}^2 R_L)/(I_{in}^2 R_{in})]$$
$$= 10 [\log_{10}(I_{out}^2/I_{in}^2) + \log_{10}(R_L/R_{out})]$$
$$= 20 \log_{10}(I_{out}/I_{in}) + 10 \log_{10}(R_L/R_{in}) \text{ decibels}$$
$$(6.3)$$

or as

$$G \text{ (or } L) = 10 \log_{10}[(V_{out}^2/R_L)/(V_{in}^2/R_{in})]$$
$$= 10 [\log_{10}(V_{out}^2/V_{in}^2) + \log_{10}(R_{in}/R_L)]$$
$$= 20 \log_{10}(V_{out}/V_{in}) + 10 \log_{10}(R_{in}/R_L) \text{ decibels}$$
$$(6.4)$$

If $R_L = R_{in}$, the term

$$10 \log_{10}(R_{in}/R_L) = 10 \log_{10}(R_L/R_{in}) = 10 \log_{10} 1 = 0$$
and then

$$G \text{ (or } L) = 20 \log_{10}(I_{out}/I_{in}) \text{ decibels} \tag{6.5}$$

or

$$G \text{ (or } L) = 20 \log_{10}(V_{out}/V_{in}) \text{ decibels} \tag{6.6}$$

Equations (6.5) and (6.6) should only be applied when the resistances in which the two currents flow, or across which the two voltages are dropped, are equal to one another. If the two resistances are not equal to one another then the basic equation

$$\text{decibels} = 10 \log_{10}(P_{out}/P_{in})$$

must be used.

Example 6.5

A network has a loss of 12 dB, an input resistance of 2000 Ω, and an output resistance of 1000 Ω. The input and output terminals are matched to the source and load resistances. Calculate (a) the current flowing in the load when the current into the network is 3 mA, and (b) the percentage error introduced into the calculation if equation (6.5) is employed.

Solution
(a) Input power $P_{in} = (3 \times 10^{-3})^2 \times 2000 = 18 \text{ mW}$.
$12 \text{ dB} = 10 \log_{10}(18/P_{out})$
$10^{1.2} = 15.85 = 18/P_{out}$
or $P_{out} = 18/15.85 = 1.14 \text{ mW}$.
Therefore $1.14 \times 10^{-3} = I_{out}^2 \times 1000$
$I_{out} = \sqrt{[(1.14 \times 10^{-3})/1000]} = 1.07 \text{ mA}$.   (*Ans.*)
(b) If the equation $20 \log_{10}(I_{out}/I_{in})$ is used then
$\qquad 12 = 20 \log_{10}(3/I_{out})$
$\qquad 10^{0.6} = 3.98 = 3/I_{out}$, and $I_{out} = 0.754 \text{ mA}$.
The percentage error $= [(0.754 - 1.07)/1.07] \times 100\%$
$= -29.5\%$ (*Ans.*)

Example 6.6

The voltage level at the output of an amplifier varies from 2.2 to 3.7 V. Express this variation in decibels.

Solution
Variation $= 20 \log_{10}(3.7/2.2) = 4.52 \text{ dB}$.   (*Ans.*)

Reference levels

The decibel is not an absolute unit like the ampere or the volt but is merely a measure of a power ratio. This means that a statement such as 'the output of an amplifier is +20 dB' is quite meaningless, since it says that the output power of the amplifier is 20 dB higher than (i.e. 100 times greater than) some unknown power. To avoid the need to state continually what the reference power is two clearly understood reference levels are often employed. These are (a) the dBm which represents decibels relative to 1 mW, and (b) the dBW which represents decibels relative to 1 W. The dBm is more commonly employed than the dBW. Thus

$$x \text{ dBm} = 10 \log_{10}[P/(1 \text{ mW})] \tag{6.7}$$

and

$$x \text{ dBW} = 10 \log_{10}(P/1) \tag{6.8}$$

where P is the power level being expressed in decibels.

Example 6.7

Express each of the following power levels in (a) dBm and (b) dBW. (i) 1 μW, (ii) 1 mW, (iii) 1 W and (iv) 1 kW.

Solution

(a)(i) 1 μW: $10 \log_{10}[(1 \times 10^{-6})/(1 \times 10^{-3})] = -30$ dBm.   (*Ans.*)

(ii) 1 mW: $10 \log_{10}[(1 \times 10^{-3})/(1 \times 10^{-3})] = 0$ dBm.   (*Ans.*)

(iii) 1 W: $10 \log_{10}[(1/(1 \times 10^{-3})] = 30$ dBm.   (*Ans.*)

(iv) 1 kW: $10 \log_{10}[(1000/(1 \times 10^{-3}] = 60$ dBm.   (*Ans.*)

(b)(i) 1 μW: $10 \log_{10}[(1 \times 10^{-6})/1] = -60$ dBW.   (*Ans.*)

(ii) 1 mW: $10 \log_{10}[(1 \times 10^{-3})/1] = -30$ dBW.   (*Ans.*)

(iii) 1 W: $10 \log_{10}(1/1) = 0$ dBW.   (*Ans.*)

(iv) 1 kW: $10 \log_{10}(1000/1) = 30$ dBW.   (*Ans.*)

Example 6.8

A tape deck has an output resistance of 1000 Ω and it is connected to a pre-amplifier whose input and output resistances are 1000 and 500 Ω respectively. In turn, the pre-amplifier supplies a main amplifier that has an input resistance of 500 Ω, an output resistance of 12 Ω, and a gain of 20 dB. The output of the main amplifier is connected to a 12 Ω loudspeaker.

If the output level of the tape deck is -12 dBm and the gain of the pre-amplifier is set so that its output power is $+20$ dBm, calculate (a) the gain of the pre-amplifier, (b) the output voltage of the pre-amplifier and (c) the power delivered to the loudspeaker.

Solution

(a) Gain $G = +20 - (-12) = 32$ dB.   (*Ans.*)

(b) $12 = 10 \log_{10}[(1 \times 10^{-3})/P]$

$\quad 10^{1.2} = 15.85 = (1 \times 10^{-3})/P$

$P = (1 \times 10^{-3})/15.85 = 63 \mu$W.

Hence, $63 \times 10^{-6} = V^2/1000$, so

$\quad V = \sqrt{(63 \times 10^{-3})} = 251$ mV.   (*Ans.*)

(c) Power to loudspeaker = 40 dBm = $10 \log_{10}[P/(1 \times 10^{-3})]$

$\quad 10^4 = P/(1 \times 10^{-3})$, and $P = 10$ W.   (*Ans.*)

Attenuation and attenuators

The attenuation, or loss, of a network is the reduction in the voltage and the current levels of a signal passing through the network that

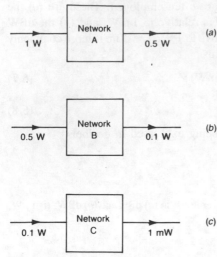

Fig. 6.5

results because of power losses within the network. The attenuation is usually quoted in decibels.

Three networks are shown in Figs. 6.5(a)–(c). When the input power to network A is 1 W the output power is only 0.5 W; the attenuation of this network is equal to $10 \log_{10}(1/0.5) = 3$ dB. Similarly, the attenuation of network B is equal to $10 \log_{10}(0.5/0.1) = 7$ dB, and the attenuation of network C is $10 \log_{10}[(0.1/(1 \times 10^{-3})] = 20$ dB. If the three networks are connected in cascade, so that the output power of network A becomes the input power to network B and the output of network B is the input to network C, the overall attenuation will be equal to the sum of the individual attenuations, i.e. $3 + 7 + 20 = 30$ dB. To confirm this, the overall attenuation $= 10 \log_{10}[1/(1 \times 10^{-3})] = 30$ dB.

A transmission line has a certain attenuation for each kilometre of line and this is known as the *attenuation coefficient* α in dB/km, of the line. The overall attenuation of a line is equal to the product of its attenuation coefficient and the length of the line in kilometres.

Example 6.9

A transmission line has an attenuation coefficient of 2.5 dB/km at a particular frequency. Calculate the attenuation of a 4 km length of this line.

Solution
Line attenuation $= 4 \times 2.5 = 10$ dB.   (*Ans.*)

Example 6.10

Determine the attenuation of the network shown in Fig. 6.6 when it is connected between a voltage source of 9 V e.m.f. and internal resistance 1500 Ω and a load resistance of 1500 Ω.

Solution
The input resistance R_{in} of the network is
$$R_{in} = 1000 + (625 \times 2500)/(625 + 2500) = 1500 \, \Omega.$$
The current flowing into the network is equal to $9/3000 = 3$ mA and the input power is $(3 \times 10^{-3})^2 \times 1500 = 13.5$ mW.
The current that flows in the load resistance is equal to
$$3 \times 625/(625 + 2500) = 0.6 \, \text{mA}.$$

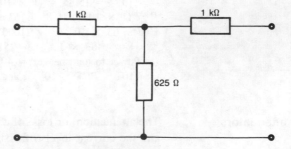

Fig. 6.6

Hence the power dissipated in the load is
$$(0.6 \times 10^{-3})^2 \times 1500 = 0.54 \, \text{mW}.$$
Therefore the attenuation L of the network is
$$10 \log_{10}(13.5/0.54) = 14 \, \text{dB}. \quad (Ans.)$$

Design of an attenuator

An attenuator can be designed to have a particular attenuation when it is connected between specified values of source and load resistance. When the source and load resistances are of the same value a symmetrical network is required and this may use either the T- or the π-configuration. Figures 6.7(a) and (b) show, respectively, a T-attenuator and a π-attenuator. When either network is connected between a specified common value of source and load resistance, as in Fig. 6.8, the input and output resistances of the network will also have this same value and a matched system is obtained. This common resistance value is known as the *characteristic resistance* R_0 of the network.

The necessary resistance values for a T-attenuator can be calculated using equations (6.9) and (6.10), or by using equations (6.11) and (6.12) for a π-attenuator.

$$\text{T-network:} \quad R_1 = R_0(N - 1)/(N + 1) \tag{6.9}$$

$$R_2 = 2R_0N/(N^2 - 1) \tag{6.10}$$

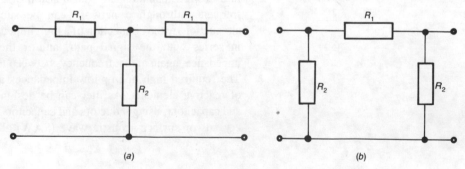

(a)   (b)

Fig. 6.7 Symmetrical attenuators:
(a) T; (b) π

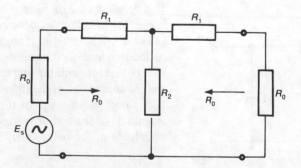

Fig. 6.8 Symmetrical T-attenuator connected between matched source and load impedances

$$\pi\text{-network: } R_1 = R_0(N^2 - 1)/2N \qquad (6.11)$$
$$R_2 = R_0(N + 1)/(N - 1) \qquad (6.12)$$

In equations $(6.9)-(6.12)$ N is the ratio

[input current (or voltage)]/[output current (or voltage)]

Example 6.11

Design (*a*) a T-attenuator and (*b*) a π-attenuator, to have 10 dB attenuation and a characteristic resistance of 250 Ω.

Solution
$20 \log_{10}N = 10$ dB, so $N = 10^{0.5} = 3.162$.
(*a*) Hence, $R_1 = (250 \times 2.162)/4.162 = 130\,\Omega$.   (*Ans.*)
$R_2 = (500 \times 3.162)/(3.162^2 - 1) = 176\,\Omega$ (*Ans.*)
(*b*) $R_1 = 250(3.162^2 - 1)/6.324 = 356\,\Omega$.   (*Ans.*)
$R_2 = (250 \times 4.162)/2.162 = 481\,\Omega$.   (*Ans.*)

If the source and load resistances are not of equal value a non-symmetrical network will have to be employed and the calculation of the required resistance values is then much more difficult. An example of a non-symmetrical L-network is given by Fig. 6.9.

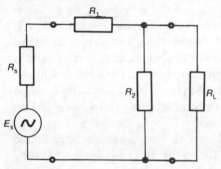

Fig. 6.9 L-attenuator

Filters

All filters are required to have an attenuation that varies with frequency in a desired manner. The transmission of a signal at an unwanted frequency through a network can be subjected to considerable attenuation by the connection of a high impedance, at that frequency, in series with the signal path, and/or the connection of a low impedance, again at that frequency, between the signal path and earth. The required high and/or low impedances are obtained by the use of reactive elements. A filter can be designed using both resistors and capacitors, using inductors and capacitors, and using piezoelectric crystals or surface-acoustic wave (SAW) elements.

First-order filter

The simplest kind of filter, known as a first-order filter, is shown in Fig. 6.10. At low frequencies the reactance of the capacitor is high and so it will have little, if any, shunting effect upon the signal path. As the frequency is increased the reactance of the capacitor will fall and so therefore will the output voltage V_{out}. This means, of course, that the attenuation of the filter increases with increase in frequency. The *cut-off frequency* f_{c} of the filter is the frequency at which the magnitude of the output voltage $|V_{\text{out}}|$ has fallen by 3 dB below the magnitude of the input voltage $|V_{\text{in}}|$.

$$|V_{\text{out}}| = |V_{\text{in}}|/\sqrt{[R^2 + (1/\omega C)^2]} \times (1/\omega C)$$

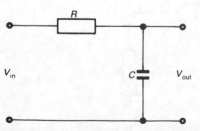

Fig. 6.10 First-order low-pass filter

$$|V_{out}/V_{in}| = 1/\sqrt{[1 + (\omega CR)^2]} \tag{6.13}$$

At the cut-off frequency f_c,

$$|V_{out}/V_{in}| = 1/\sqrt{2} = 1/\sqrt{[1 + (\omega_c CR)^2]}$$

or

$$\omega_c = 1/CR$$

and

$$f_c = 1/2\pi CR \tag{6.14}$$

Example 6.12

A filter of the type shown in Fig. 6.10 has $C = 0.1\ \mu F$ and $R = 10\ k\Omega$. Calculate the frequencies at which the output voltage is (a) 3 dB, (b) 6 dB and (c) 12 dB down on the input voltage. Plot the attenuation/frequency characteristic of the filter.

Solution
(a) $f_c = 1/2\pi CR = 1/[2\pi(10 \times 10^3 \times 0.1 \times 10^{-6})] = 159$ Hz.   (*Ans.*)
(b) 6 dB = voltage ratio of 2 and hence
$$2 = \sqrt{[1 + (\omega_6 CR)^2]}$$
$$4 = 1 + (\omega_6 CR)^2$$
$$\omega_6 = \sqrt{3}/CR = \sqrt{3} \times 1000. \text{ Therefore}$$
$$f_6 = (\sqrt{3} \times 1000)/2\pi = 275.7 \text{ Hz.}\quad (\textit{Ans.}).$$
(c) 12 dB = voltage ratio of 3.98 $\simeq$ 4. Therefore
$$16 = 1 + (\omega_{12} CR)^2$$
$$\omega_{12} = \sqrt{15}/CR \text{ and}$$
$$f_{12} = (\sqrt{15} \times 1000/2\pi) = 616.4 \text{ Hz.}\quad (\textit{Ans.})$$
The attenuation/frequency characteristic of the filter is shown plotted by Fig. 6.11.

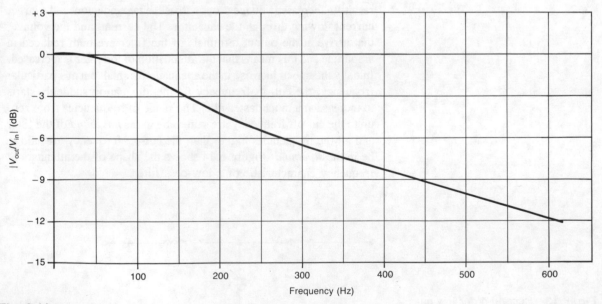

Fig. 6.11

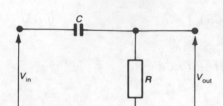

Fig. 6.12 First-order high-pass filter

A first-order high-pass filter can be obtained by interchanging the positions of the resistor and the capacitor in the low-pass filter to give the circuit shown in Fig. 6.12.

For many applications the *roll-off* of the first-order filter is not rapid enough and then a higher-order filter must be employed. A second-order filter can be obtained by the use of both inductance and capacitance or by the use of an *active filter*.

Inductor–capacitor filters

An inductor–capacitor filter uses both inductance and capacitance to produce the required attenuation/frequency characteristic. The inductors possess inevitable self-resistance and this ensures that there is some loss in the passband.

The low-pass filter

The circuit of the basic low-pass filter is shown in Fig. 6.13. If a variable frequency, constant voltage source is applied to the input terminals of the filter the output voltage V_{out} will vary with frequency. At low frequencies the reactances of the two inductors will be low and the reactance of the capacitor will be high; at these frequencies therefore little voltage is dropped across the inductors and only a small current flows through the capacitor. Most of the signal voltage applied to the input terminals of the filter appears at the output terminals and so the attenuation of the filter is small. As the frequency of the voltage source is increased the reactances of the two inductors increase and the reactance of the capacitor falls; this results in increased voltage drops across the two inductors and a larger current flowing through the capacitor. The current, and the voltage, that arrive at the output terminals of the filter are both reduced in amplitude and this means that the attenuation of the filter is increased. Initially the rate of increase in the attenuation is small but at a particular frequency, the cut-off frequency f_c, the attenuation suddenly starts to increase at a much faster rate. The range of frequencies from zero up to the cut-off frequency are said to be in the *passband* of the filter and all the frequencies higher than the cut-off frequency are said to be in the *stopband*. Figure 6.14 shows the shape of the attenuation/frequency characteristic of a low-pass filter.

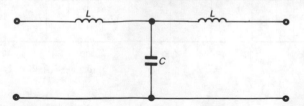

Fig. 6.13 Second-order low-pass filter

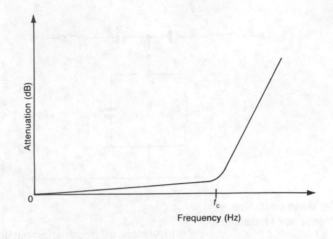

Fig. 6.14 Attenuation/frequency characteristic of second-order low-pass filter

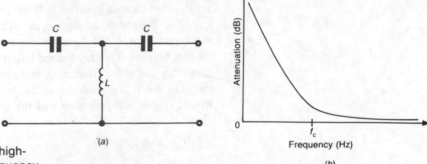

Fig. 6.15 (a) Second-order high-pass filter; (b) attenuation/frequency characteristic

High-pass filter

The basic high-pass filter is shown by Fig. 6.15(a) and its attenuation/frequency characteristic by Fig. 6.15(b). At low frequencies the reactances of the two capacitors are high and the reactance of the inductance is low. There is then a relatively large voltage drop across each of the capacitors and a relatively large current flowing through the inductor. Little current and voltage, and hence power, is able to reach the output terminals of the filter and so its attenuation is high. With increase in frequency the reactances of the two capacitors fall and the inductor's reactance rises and so the attenuation of the filter falls rapidly. At a particular frequency, the cut-off frequency f_c, the rate of decrease of the attenuation becomes small and thereafter falls only little with further increase in frequency. The stopband of the filter is from zero hertz to the cut-off frequency and the passband contains all those frequencies that are higher than the cut-off frequency.

Band-pass filter

A band-pass filter is required to pass a specified band of frequencies with the minimum attenuation and to provide considerable attenuation

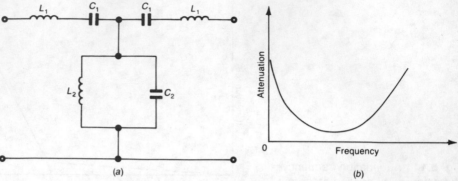

Fig. 6.16 (a) Band-pass filter; (b) attenuation/frequency characteristic

to signals at frequencies outside of this bandwidth. The desired attenuation/frequency characteristic can be obtained by the use of series-tuned circuits in series with the signal path and a parallel-tuned circuit in shunt with the signal path. The circuit of a band-pass filter is shown by Fig. 6.16(a) and its attenuation /frequency characteristic by Fig. 6.16(b). The three tuned circuits are all resonant at the same frequency and at this frequency, and a band of frequencies either side, the filter offers its minimum attenuation, the only loss being due to the inevitable self-resistances of the inductors.

Band-stop filter

The fourth kind of filter is known as the band-stop filter, and its function is to offer a large attenuation to a specified band of frequencies and to offer the minimum attenuation to all frequencies lying outside of this bandwidth. The circuit of a band-stop filter is shown by Fig. 6.17(a) and Fig. 6.17(b) shows its attenuation/frequency characteristic.

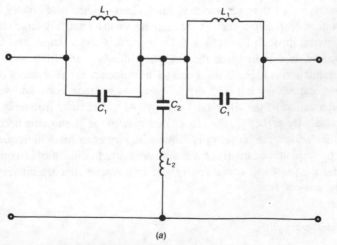

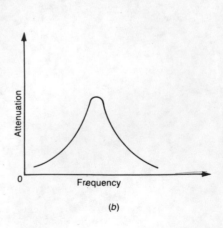

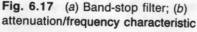

Fig. 6.17 (a) Band-stop filter; (b) attenuation/frequency characteristic

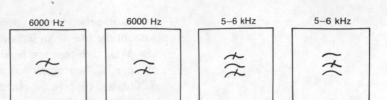

Fig. 6.18

Example 6.13

The band of frequencies 1000–12 000 Hz is applied to each of the filters shown in Fig. 6.18. Determine the frequencies at the output of each filter.

Solution
(*a*) Low-pass filter: 1–6 kHz.   (*Ans.*)
(*b*) High-pass filter: 6–12 kHz.   (*Ans.*)
(*c*) Band-pass filter: 5–6 kHz.   (*Ans.*)
(*d*) Band-stop filter: 1–5, and 6–12 kHz.   (*Ans.*)

Active filters

Inductors are relatively bulky components, particularly at the values required at the lower frequencies, and they have inherent losses that are difficult to predict accurately. The use of an op-amp and resistor–

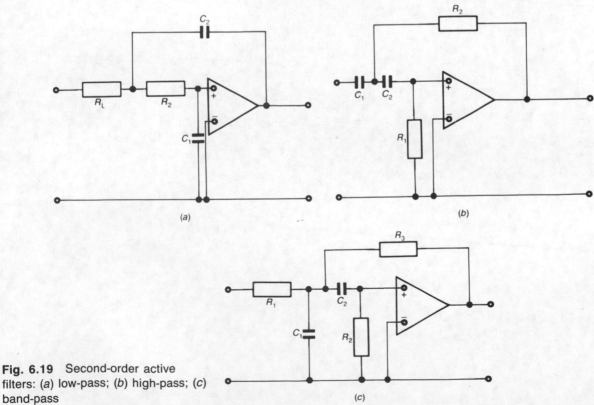

Fig. 6.19 Second-order active filters: (*a*) low-pass; (*b*) high-pass; (*c*) band-pass

capacitor network will allow a second-order filter to be obtained without the use of an inductor. The circuits of second-order active low-pass, high-pass and band-pass filters are given in Fig. 6.19. The op-amp is often employed to provide a first-order filter. Figures 6.20(a) and (b) give the circuits for a low-pass and a high-pass active filter respectively.

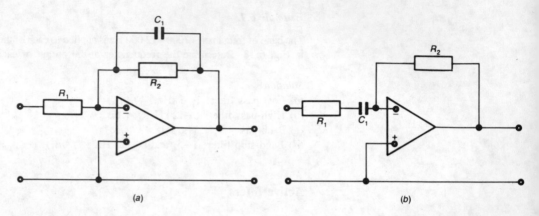

Fig. 6.20 First-order active filters:
(a) low-pass; (b) high-pass

7 Modulation

Modulation is the process of superimposing information on to a carrier wave. One of the characteristics of the carrier wave must be varied by the modulating or *baseband* signal. If the carrier is a sinusoidal wave then the characteristics that may be varied are its amplitude, giving *amplitude modulation* (AM), its frequency, giving *frequency modulation* (FM), or its phase, to give *phase modulation*. The modulating signal may be either an analogue or a digital signal. If the carrier is a rectangular pulse train then the amplitude, or the width (duration), or the position in time of each pulse may be varied by the baseband signal to give, respectively, *pulse amplitude modulation* (PAM), *pulse duration modulation* (PDM) or *pulse position modulation* (PPM). Information about the height of each pulse in a PAM waveform may be encoded, using the binary code, to produce *pulse code modulation* (PCM).

Some kind of modulation is always employed in a communication system to enable the best use to be made of the bandwidth provided by the transmission medium.

Firstly, the baseband signal may not be able to propagate over the transmission medium.

(a) Digital data signals cannot propagate over the public switched telephone network (PSTN) because the PSTN cannot transmit frequencies that are either below 300 Hz or above 3400 Hz. Modulation must therefore be employed to convert the data signals from their digital form into voice-frequency (VF) form so that they can travel over the PSTN in the same way as speech signals. At the receiving end of a data link the reverse process, known as *demodulation*, must be employed to convert the VF signals back to their original digital form.

(b) Aerials can only operate at frequencies higher than about 15 kHz and this means that all baseband signals that are to be transmitted over a radio link must first modulate an RF carrier in order to be shifted to another part of the frequency spectrum. Different signals must be shifted to different radio frequencies so that the radio receiver will be able to select the wanted signal from all those signals that are simultaneously present at the receive aerial. For example, the BBC sound broadcast services use the frequencies given in Table 7.1.

Table 7.1

Station	Frequencies	Station	Frequencies
Radio 1	97.6–99.8 MHz. 1053 and 1089 kHz.	Radio 4	92.4–94.6 MHz. 198 kHz.
Radio 2	88–90.2 MHz	Radio 5	693 and 909 kHz.
Radio 3	90.2–92.4 MHz. 1215 kHz.		

Secondly, to increase the channel-carrying capacity of a transmission medium either *frequency-division multiplex* (FDM), or *time-division multiplex* (TDM) may be employed. Frequency-division multiplex employs either amplitude, or frequency, modulation to position different channels at various positions in the bandwidth made available by the transmission medium. It finds its main applications in some multi-channel radio systems and in some data systems. Time-division multiplex systems allow a number of channels to share the transmission medium on a sequential time basis. Each channel has the exclusive use of the medium given at regular intervals for a short period of time.

Amplitude modulation

Amplitude modulation is the process whereby the amplitude of a sinusoidal carrier wave is varied by a modulating signal. The modulating signal and the carrier are both applied to a circuit known as a *modulator* as shown by Fig. 7.1. The amplitude of the carrier is made to vary in direct proportion to the amplitude of the modulating signal a number of times per second equal to the frequency of the modulating signal. Neither the frequency nor the phase of the carrier are altered. The AM wave is transmitted over the transmission medium to the distant end of the link. Here the wave must be *demodulated*, or *detected*, to recover the original modulating signal. The demodulation process is carried out by a circuit known as a demodulator, or a detector. Some demodulators require the unmodulated carrier wave to be reinserted, as shown by the dotted line, before demodulation can take place while other circuits do not have this requirement.

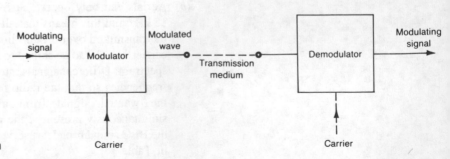

Fig. 7.1 Basic AM system

Figure 7.2(*a*) shows a sinusoidal modulating signal of amplitude V_m and frequency f_m, and Fig. 7.2(*b*) shows a carrier wave of amplitude V_c and frequency f_c. When the carrier wave is amplitude modulated by the modulating signal the amplitude of the modulated wave, shown by Fig. 7.2(*c*), varies sinusoidally at frequency f_m. The maximum value of the modulated wave is equal to $V_c + V_m$ and its minimum value is $V_c - V_m$. The outline of the modulated carrier waveform, shown dotted, is known as the *modulation envelope*. Clearly, both the positive and the negative sides of the modulation envelope have the same waveform as the modulating signal.

The amplitude of the modulated wave is equal to $V_c + V_m \sin \omega_m t$, and so the instantaneous voltage of the AM carrier wave is given by

$$v = (V_c + V_m \sin \omega_m t) \sin \omega_c t \text{ V} \qquad (7.1)$$

For a signal to carry information it must contain at least two

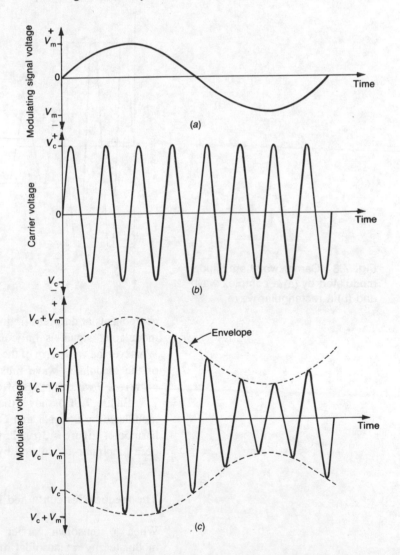

Fig. 7.2 (*a*) Sinusoidal modulating signal; (*b*) carrier wave; (*c*) AM wave

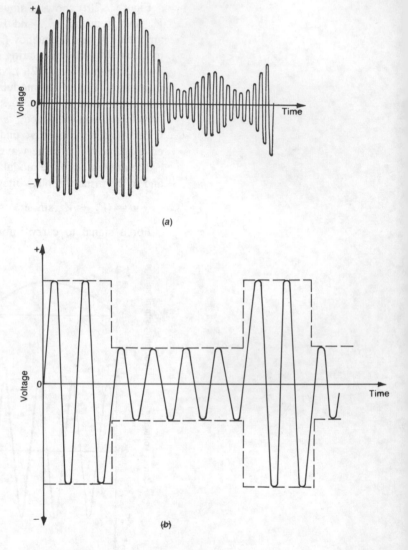

Fig. 7.3 Carrier wave amplitude modulated by (a) a complex wave and (b) a rectangular wave

components at different frequencies and so, in practice, a sinusoidal modulating signal is only used for test or monitoring purposes. Whatever the waveform of the complex modulating signal the envelope of the modulated wave must be identical or distortion will have occurred. Two examples of complex waveforms are given in Fig. 7.3. Figure 7.3(a) shows the waveform of a carrier that has been modulated by a signal that consists of a fundamental plus its third harmonic. Figure 7.3(b) shows the waveform of a carrier that has been amplitude modulated by a rectangular modulating signal.

The frequencies contained in an amplitude-modulated wave

When a sinusoidal carrier wave of frequency f_c is amplitude modulated by a sinusoidal modulating signal, at frequency f_m, the

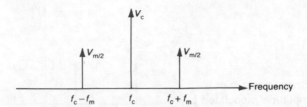

Fig. 7.4 Spectrum diagram of a sinusoidally modulated carrier wave

modulated wave will contain components at three different frequencies (see Appendix A). These three frequencies are:

(a) the carrier frequency f_c;
(b) the lower side frequency $f_c - f_m$;
(c) the upper side frequency $f_c + f_m$.

The amplitude of the carrier component is the same as its unmodulated value V_c, and the amplitude of each of the two side frequencies is equal to one-half of the modulating signal voltage. The AM wave does not contain a component at the modulating signal frequency f_m. The spectrum diagram of a sinusoidally modulated carrier wave is given by Fig. 7.4. The length of each arrow is made directly proportional to the voltage of the component that it represents.

The minimum bandwidth B that is necessary to transmit the modulated waveform is equal to the difference between the highest frequency and the lowest frequency contained in the modulated wave. Hence

$$B = (f_c + f_m) - (f_c - f_m) = 2f_m \tag{7.2}$$

Clearly, the minimum bandwidth required is equal to twice the frequency of the modulating signal.

Example 7.1

A 100 kHz carrier wave is amplitude modulated by a sinusoidal modulating signal at 10 kHz. Calculate the frequencies contained in the modulated waveform and the minimum bandwidth that is necessary for its transmission.

Solution
The frequencies contained in the AM wave are:
(a) The carrier frequency $f_c = 100$ kHz.   (*Ans.*)
(b) The lower side frequency $f_c - f_m = 100 - 10 = 90$ kHz.   (*Ans.*)
(c) The upper side frequency $f_c + f_m = 100 + 10 = 110$ kHz.   (*Ans.*)
The minimum bandwidth that is required $= 110 - 90 = 20$ kHz.   (*Ans.*)

When the modulating signal is complex it will contain components at two, or more, frequencies. If the highest frequency in the modulating signal is f_2 then it will give rise to lower and upper side frequencies $f_c \pm f_2$ in the modulated wave. Similarly, if the lowest modulating frequency is f_1 lower and upper side frequencies $f_c \pm f_1$ will be produced. This means that there will be a band of upper side frequencies ranging from $f_c + f_1$ to $f_c + f_2$ lying above the carrier

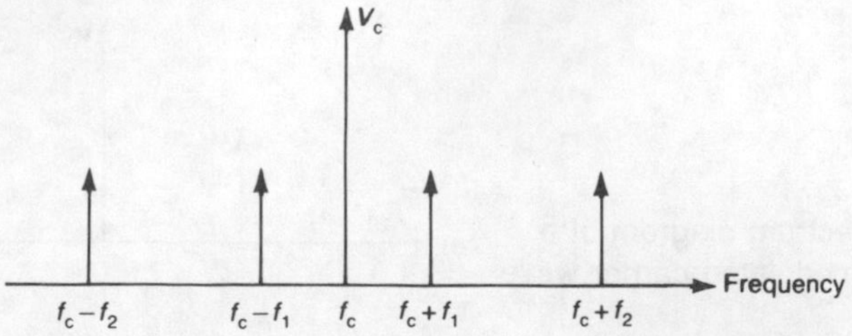

Fig. 7.5 Spectrum diagram of a carrier wave amplitude modulated by a complex signal

frequency. This band of frequencies is known as the *upper sideband*. Similarly, there will also be a *lower sideband* below the carrier frequency that contains components at frequencies in the band $f_c - f_2$ to $f_c - f_1$. In general, if there are n components at different frequencies in the modulating signal there will be $2n$ side frequencies in the modulated wave. The spectrum diagram is shown by Fig. 7.5. The two sidebands are symmetrically situated either side of the carrier frequency. The lower sideband is said to be *inverted* because the highest frequency in it, i.e. $f_c - f_1$, corresponds to the lowest frequency in the modulating signal and vice versa. The upper sideband is said to be *erect* since the highest frequency it contains, i.e. $f_c + f_2$, corresponds to the highest modulating frequency. The minimum bandwidth that is needed to transmit the modulated signal is equal to twice the highest frequency that is contained in the modulating signal.

Example 7.2

A 7 MHz carrier is amplitude modulated by the band of frequencies 50 Hz to 5 kHz. Calculate (*a*) the frequencies contained in the lower and upper sidebands and (*b*) the minimum bandwidth required.

Solution
(*a*) Lower sideband = 7 MHz − (50 Hz to 5 kHz)
$\qquad\qquad\qquad$ = 6.995–6.99995 MHz.   (*Ans.*)
$\quad$ Upper sideband = 7 MHz + (50 Hz to 5 kHz)
$\qquad\qquad\qquad$ = 7.00005–7.005 MHz.   (*Ans.*)
(*b*) The minimum bandwidth = 2 × 5 kHz = 10 kHz.   (*Ans.*)

Modulation factor

The *modulation factor m* of an AM wave is a means of expressing the amount of modulation:

$$m = \text{(maximum amplitude − minimum amplitude)/(maximum amplitude + minimum amplitude)} \qquad (7.3)$$

When the modulation factor is expressed as a percentage it is often known as the depth of modulation.

For a sinusoidally modulated wave the maximum amplitude of the modulated waveform is $V_c + V_m$ and the minimum amplitude is $V_c - V_m$. Therefore

$$m = [(V_c + V_m) - (V_c - V_m)]/[(V_c + V_m) + (V_c - V_m)]$$
$$= V_m/V_c \tag{7.4}$$

The expression for the instantaneous voltage of an AM wave can now be rewritten in terms of the modulation factor:

$$v = V_c[1 + (V_m/V_c) \sin \omega_m t] \sin \omega_c t$$
$$= V_c[1 + m \sin \omega_m t] \sin \omega_c t \tag{7.5}$$

Example 7.3

The envelope of an AM wave varies between a maximum value of ± 55 V and a minimum value of ± 30 V. Calculate (a) the carrier voltage, (b) the modulating signal voltage, (c) the voltage of each side frequency and (d) the modulation factor.

Solution
(a) $55 = V_c + V_m$; $30 = V_c - V_m$. Adding
 $\quad 85 = 2V_c$, or $V_c = 42.5$ V.   (*Ans.*)
(b) $V_m = 42.5 - 30 = 12.5$ V.   (*Ans.*)
(c) $V_{SF} = V_m/2 = 6.25$ V.   (*Ans.*)
(d) $m = V_m/V_c = 12.5/42.5 = 0.294$.   (*Ans.*)

Example 7.4

The instantaneous voltage of an AM wave is
 $\quad v = 100 (1 + 0.6 \sin 4000\pi t) \sin (40\pi \times 10^6 t)$ volts.
Calculate (a) the carrier voltage, (b) the modulating signal voltage, (c) the carrier frequency and (d) the modulating signal frequency. Draw the modulated waveform.

Solution
(a) $V_c = 100$ V.   (*Ans.*)
(b) $V_m = mV_c = 0.6 \times 100 = 60$ V.   (*Ans.*)
(c) $f_c = (40\pi \times 10^6)/2\pi = 20$ MHz.   (*Ans.*)
(d) $f_m = 4000\pi/2\pi = 2$ kHz.   (*Ans.*)
The modulation envelope varies between a maximum value of 160 V and a minimum value of 40 V. The waveform is shown in Fig. 7.6.

Power in an amplitude-modulated wave

When an AM wave is applied across a resistance R power will be dissipated by each of the components in the wave. For a sinusoidally modulated wave:

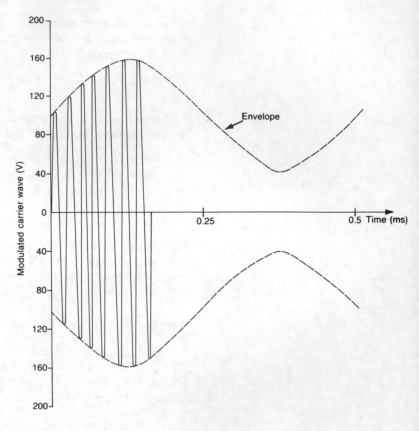

Fig. 7.6

(a) The carrier component has an r.m.s. voltage of $V_c/\sqrt{2}$ and it dissipates a power of $V_c^2/2R$.

(b) The lower side frequency component has a peak voltage of $V_m/2 = mV_c/2$. The r.m.s. voltage is $(m_cV_c)/2\sqrt{2}$ and the power dissipated is $m^2V_c^2/8R$.

(c) The upper side frequency component has the same voltage as the lower side frequency component and so it also dissipates a power of $m^2V_c^2/8R$.

The total power P_T dissipated is equal to the sum of the three individual powers. Therefore

$$P_T = V_c^2/2R + m^2V_c^2/4R$$
$$= (V_c^2/2R)(1 + m^2/2)$$

or

$$P_T = P_c(1 + m^2/2) \quad \text{W} \tag{7.6}$$

where P_c is the power dissipated by the carrier component.

Example 7.5

A 1 kW carrier wave is amplitude modulated to a depth of 70%. Calculate (a) the total power and (b) the power in the side frequencies.

Solution
(*a*) Total power = 1000 $(1 + 0.7^2/2)$ = 1245 W.   (*Ans.*)
(*b*) Side frequency power = 1245 − 1000 = 245 W.   (*Ans.*)

Example 7.5 shows that the power contained in the two side frequencies is only a small fraction of the total power (actually 19.7%). The maximum side frequency power occurs when the modulation factor is equal to unity and even then it is only 33.3% of the total power. This means that AM is not a very efficient system when considered on a power basis and so *single-sideband* (SSB) operation is often employed for radio systems. An SSB transmitter suppresses both the carrier component and one of the sidebands to obtain both increased efficiency and a reduced bandwidth.

Overmodulation

The theoretical maximum depth of modulation is 100% when $m = 1$ and the modulation envelope varies between a minimum value of $V_c(1 - m) = 0$ and a maximum value of $V_c(1 + m) = 2V_c$. If the depth of modulation is still further increased the modulation envelope is unable to follow the waveform of the modulating signal as shown by Fig. 7.7. Considerable envelope distortion will then have occurred and so overmodulation is never used in practice.

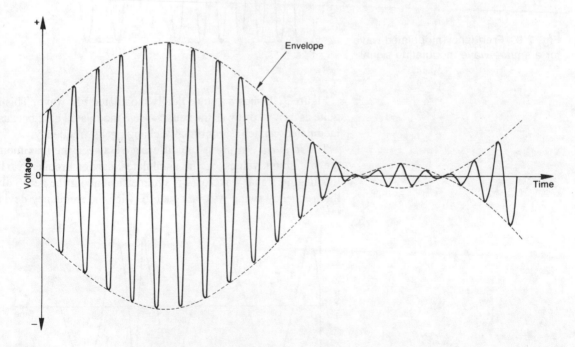

Fig. 7.7 Amplitude-modulated wave with a modulation depth in excess of 100%

Frequency modulation

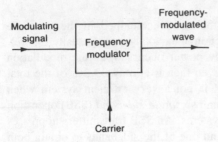

Fig. 7.8 Frequency modulation

When a sinusoidal carrier wave is frequency modulated the instantaneous carrier frequency is deviated either side of its mean, unmodulated value. The frequency deviation is directly proportional to the amplitude of the modulating signal, and the number of times per second that the carrier frequency is deviated is equal to the modulating frequency. The amplitude of the carrier is kept constant. Frequency modulation is achieved by applying both the modulating signal and the carrier to a circuit known as a *frequency modulator* (see Fig. 7.8). The frequency deviation of the carrier is equal to the *sensitivity* of the modulator in kHz/V times the modulating signal voltage. Consider a simple example. Suppose that a 2 V peak, 5 kHz carrier is frequency modulated by a 500 Hz ± 0.5 V square wave and the frequency deviation is ± 1 kHz. The sensitivity of the modulator is equal to $1000/0.5 = 2$ kHz/V. When the modulating signal voltage is + 0.5 V the carrier frequency is equal to $5 + 1 = 6$ kHz; when the modulating signal voltage is − 0.5 V the carrier frequency is $5 − 1 = 4$ kHz. The FM waveform is shown in Fig. 7.9.

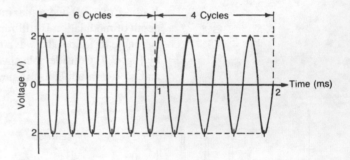

Fig. 7.9 Frequency-modulated wave for a square-wave modulating signal

Figure 7.10 shows a sinusoidally modulated FM wave. This is not quite as easy to draw as the square-wave modulated case because the modulated carrier is continually changing its frequency.

The *frequency swing* is the difference between the maximum and the minimum values of the modulated carrier frequency. If the modulating signal waveform is symmetrical about the zero volt axis the frequency swing will be equal to twice the frequency deviation.

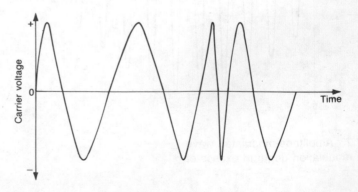

Fig. 7.10 Frequency-modulated wave for a sinusoidal modulating signal

Modulation index

The *modulation index* m_f of an FM wave is

$$m_f = \text{(frequency deviation)/(modulating signal frequency)} \qquad (7.7)$$

and it is equal to the peak *phase* deviation of the carrier in radians.

Example 7.6

A 100 MHz carrier wave is frequency modulated by a 10 kHz signal. If the peak frequency deviation is 60 kHz calculate (*a*) the maximum and minimum carrier frequencies, (*b*) the frequency swing and (*c*) the modulation index.

Solution
(*a*) $f_{max} = 100\,\text{MHz} + 60\,\text{kHz} = 100.06\,\text{MHz}.$ (*Ans.*)
     $f_{min} = 100\,\text{MHz} - 60\,\text{kHz} = 99.94\,\text{MHz}.$ (*Ans.*)
(*b*) Frequency swing $= 2 \times 60 = 120\,\text{kHz}.$ (*Ans.*)
(*c*) $m_f = 60/10 = 6\,\text{rad}.$ (*Ans.*)

The bandwidth occupied by an FM wave is larger than the frequency swing and it increases with increase in the frequency deviation. This means that an arbitrary limit must be placed upon the maximum frequency deviation permitted in a particular FM system. This maximum is known as the *rated system deviation*.

Example 7.7

A frequency modulator has a sensitivity of 20 kHz/V. If the rated system deviation is 75 kHz what is the maximum allowable modulating signal voltage?

Solution
Maximum modulating signal voltage $= 75/20 = 3.75\,\text{V}.$ (*Ans.*)

Deviation ratio

The *deviation ratio D* of an FM system is the particular case of the modulation index when both the frequency deviation (and hence the modulating signal voltage), and the modulating signal frequency are at their maximum permissible values. The deviation ratio is used in the design of an FM system. Therefore

$$D = \text{(maximum frequency deviation)/(maximum modulating frequency)} \qquad (7.8)$$

The frequencies contained in a frequency-modulated wave

When a carrier at frequency f_c is frequency modulated by a sinusoidal modulating signal at frequency f_m a number of side frequencies may be generated. In general, the modulated wave will contain components at the following frequencies:

(a) the carrier frequency f_c;
(b) the first-order side frequencies $f_c \pm f_m$;
(c) the second-order side frequencies $f_c \pm 2f_m$;
(d) the third-order side frequencies $f_c \pm 3f_m$;
(e) the nth-order side frequencies $f_c \pm nf_m$m.

The number of these possible side frequencies that may exist in any particular FM wave is determined by the value of the modulating index.

(a) *Narrow-band frequency modulation* (NBFM). If the modulation index is less than unity ($m_f < 1$), only the carrier frequency and the first-order side frequencies will be present in the modulated wave. The components contained in the wave are then the same as if the carrier had been amplitude modulated. NBFM is used for mobile radio services where each channel is allocated a very narrow bandwidth.

(b) *Wideband frequency modulation*. As the modulation index is increased above unity, higher and higher orders of side frequencies are generated, e.g. when $m_f = 5$ up to the eighth-order of side frequency is present in the modulated wave. In addition any component, including the carrier, may be of zero amplitude at any particular value of m_f, e.g. the carrier is zero when $m_f = 2.405$ and 5.32.

Bandwidth

The bandwidth B required for the satisfactory transmission of an FM wave is given by

$$B = 2(f_d + f_m) \tag{7.9}$$

where f_d = maximum frequency deviation and f_m = maximum modulating signal frequency. Equation (7.9) assumes that the highest-order side frequencies whose voltage is less than 1% of the unmodulated carrier voltage need not be transmitted.

Example 7.8

A 2 W, 100 MHz carrier is modulated first in amplitude and then in frequency by a 5 kHz sinusoidal signal. The resulting values of the modulation factor and modulation index are both 0.5. Calculate (a) the bandwidth occupied by each modulated wave, (b) the power contained in each modulated wave and (c) the new occupied bandwidths if the modulating frequency were reduced to 500 Hz with the modulating voltage remaining unchanged.

Solution

(a) AM: $B = 2f_m = 10$ kHz.   (*Ans.*)

　　FM: $0.5 = f_d/5000$, or $f_d = 2500$ Hz.

　　　　$B = 2(2500 + 5000) = 15$ kHz.   (*Ans.*)

(b) AM: $P_T = 2(1 + 0.5^2/2) = 2.25$ W.   (*Ans.*)

　　FM: $P_T = P_c = 2$ W.   (*Ans.*)

(c) AM: $B = 2f_m = 1000$ Hz.   (*Ans.*)

　　FM: $B = 2(2500 + 500) = 6$ kHz.   (*Ans.*)

Demodulation

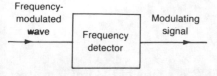

Frequency-modulated wave → Frequency detector → Modulating signal

Fig. 7.11 Demodulation of an FM wave

The demodulation of an FM wave is the reverse process to modulation and it consists of the recovery of the information contained in the modulated wave. The basic concept is illustrated by Fig. 7.11. Most of the FM detectors that use discrete components first convert the FM signal into an AM signal and then amplitude detect this to obtain the baseband signal. Integrated circuit FM detectors generally employ either a phase-locked loop or a quadrature detector.

The relative merits of amplitude and frequency modulation

Frequency modulation of a carrier requires the provision of a wider bandwidth than does AM. This bandwidth may be only a little wider, as in the case of NBFM, or very much wider as in the case, for example, of VHF sound broadcast signals. These have a rated system deviation of 75 kHz and a maximum modulating frequency of 15 kHz and so the minimum bandwidth to be provided is $2(75 + 15) = 180$ kHz. This should be compared with the 30 kHz that would be required by an AM system transmitting the same signal.

When a carrier is amplitude modulated its r.m.s. voltage, and hence its power, increases. This is not so for an FM wave whose voltage, and hence power, remains constant at its unmodulated value. This feature results in FM transmitters being more efficient than AM transmitters with consequent economies in power consumption.

Another advantage of frequency modulation is with regard to *signal-to-noise ratio*. Signal-to-noise ratio is the ratio (signal power)/(noise power) and it is a very important factor in all communication systems. An FM system gives a better output signal-to-noise ratio than an AM system, the improvement being given by

$$\text{Signal-to-noise ratio improvement} = 20 \log_{10}[\sqrt{(3)}D] \text{ decibels} \tag{7.10}$$

Example 7.9

An AM system has an output signal-to-noise ratio of 40 dB. What would be the new output signal-to-noise ratio if the system were changed to FM with a deviation ratio of $\sqrt{3}$?

Solution

FM output signal-to-noise ratio $= 40 + 20 \log_{10}(\sqrt{3} \sqrt{3})$
$$= 49.54 \, \text{dB.} \quad (\textit{Ans.})$$

Amplitude modulation is used for long, medium and short waveband sound broadcasting, for the picture signal in television broadcasting, for some VHF/UHF mobile radio systems, and for various aero and maritime applications. Frequency modulation is used for VHF sound broadcasts, for the sound signal of television broadcasts, for some VHF/UHF mobile radio systems, for some aero/maritime applications, and for wideband radio-relay systems.

Frequency-shift modulation

Frequency-shift modulation (FSK) is a form of FM that is used for low-speed data circuits. The nominal carrier frequency is always deviated to either one of two frequencies; the higher frequency f_0 represents binary 0 and the lower frequency f_1 represents binary 1.

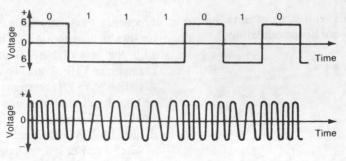

Fig. 7.12 Frequency-shift modulation signal

Table 7.2

Bit rate (b/s)	Frequencies		
	f_0	f_1	
Up to 300	1180	980	Different
	1850	1650	directions
600	1700	1300	
1200	2100	1300	

An example of an FSK signal is shown in Fig. 7.12. The choice of the two frequencies is based upon the available bandwidth (severely limited by the PSTN), and the required bit rate. The higher the bit rate the larger must be the spacing between the two frequencies so that the receiver will be able to discriminate between them. In practice, this means that FSK can only be employed for bit rates of up to 1200 b/s. The bit rates and the frequencies employed have been specified by the CCITT† and they are shown in Table 7.2.

---

†CCITT: Consultative Committee International for Telephony and Telegraphy.

From equation (7.9) the minimum bandwidth required for an FM wave is $2(f_d + f_m)$. For an FSK system $2f_d = f_0 - f_1$ and $f_{m(max)}$ = (bit rate)/2, and hence the necessary bandwidth B is apparently

$$B = 2[(f_0 - f_1)/2 + \text{(bit rate)}/2]$$

or

$$B = f_0 - f_1 + \text{bit rate}. \tag{7.11}$$

Example 7.10

Calculate the bandwidth occupied by an FSK signal if the bit rate is (a) 600 b/s and (b) 1200 b/s.

Solution
(a) From Table 7.2,
  $B = (1700 - 1300) + 600 = 1000\,\text{Hz}.$ *(Ans.)*
(b) $B = (2100 - 1300) + 1200 = 2000\,\text{Hz}.$ *(Ans.)*

Unfortunately, the 2 kHz bandwidth required by the 1200 b/s FSK signal is not available when the PSTN is used as the transmission medium. (The available bandwidth is from 900 to 2100 Hz or 1200 Hz.) However, enough information is contained in the first-order side frequencies for the data receiver to be able to determine, at each instant, whether a binary 0 or a binary 1 bit is being received and so correctly receive the signal. If only the first-order side frequencies are transmitted the minimum bandwidth B becomes equal to the bit rate, i.e.

$$B = \text{bit rate} \tag{7.12}$$

Phase modulation

Phase modulation of a carrier occurs when the phase of the carrier is varied directly in proportion to the amplitude of the modulating signal voltage. The number of times per second the phase is deviated is equal to the frequency of the modulating signal. Phase modulation finds some application in analogue VHF/UHF mobile systems but its main applications are digital radio systems and some data communication systems. The basic principle of phase shift modulation (PSK) is illustrated by Fig. 7.13. The phase of the carrier is shifted by 180° each time that the digital signal changes from binary 1 to binary 0, or from binary 0 to binary 1. The demodulation of a PSK signal presents some difficulties, because a reference phase must be made available at the receiver, and therefore a version of it, known as *differential phase shift modulation* (DPSK) is employed in data systems. With DPSK changes in phase, rather than phase itself, are used to represent values and these changes are easier for the receiver to detect. To reduce the transmission rate over the line the data stream is split into groups of either two bits (dibits), or three bits (tribits). Dibits represent the four states 00, 01, 11 and 10, and tribits represent

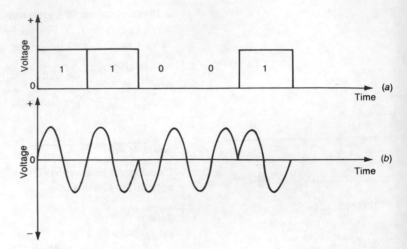

Fig. 7.13 Phase modulated wave

the eight states 000 through to 111. The DPSK version of phase modulation is used at bit rates of from 1200 to 4800 b/s.

Higher bit rates, from 2400 to 19,800 b/s, use *quadrature amplitude modulation* (QAM). This is a mixture of AM and phase modulation and it is used in conjunction with quabits, which are groups of four bits.

Pulse modulation

In a pulse modulation system the baseband signal is sampled at regular intervals of time and a pulse is transmitted to line to represent the voltage of each sample. So that information about different signal amplitudes can be signalled a characteristic of each pulse must be modulated. Four main kinds of pulse modulation exist: pulse amplitude modulation, pulse duration modulation, pulse position modulation and pulse code modulation. The principle of each of the first three kinds of pulse modulation is illustrated by Fig. 7.14.

Pulse amplitude modulation

Figure 7.14(*a*) shows a pulse amplitude modulated (PAM) waveform in which equal width pulses having a fixed time interval between their leading edges have their amplitudes modulated by the baseband signal. When the baseband signal is positive the amplitude of a pulse is greater than its mean, unmodulated, value; conversely, whenever the baseband signal is negative the pulse amplitude is less than its unmodulated value. A PAM waveform contains components at a number of different frequencies:

(*a*) the baseband frequencies;

(*b*) the pulse repetition frequency (PRF) and AM lower and upper sidebands centred on the PRF;

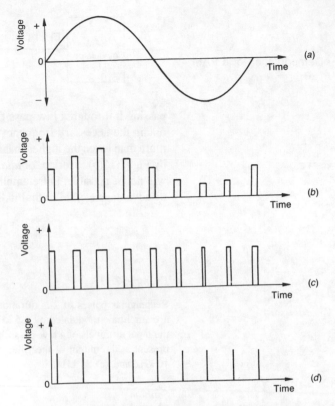

Fig. 7.14 Methods of pulse modulation: (*a*) modulating signal; (*b*) PAM; (*c*) PDM; (*d*) PPM

(*c*) higher frequencies than (*b*) such as AM sidebands centred on integer multiples of the PRF;

(*d*) a d.c. component whose voltage is equal to the mean value of the PAM waveform.

The spectrum diagram of a PAM waveform is shown in Fig. 7.15. If there is a gap between the highest frequency in the baseband signal and lowest frequency in the lower sideband of the PRF, as in Fig. 7.15(*a*), it will then be possible to demodulate the PAM signal by

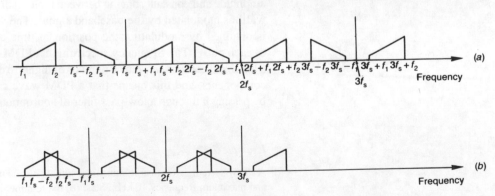

Fig. 7.15 Spectrum diagram of a PAM wave

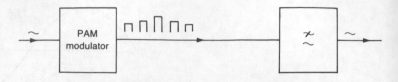

Fig. 7.16 Demodulation of a PAM signal

passing it through a low-pass filter. This is shown by Fig. 7.16. To obtain the necessary frequency gap the sampling frequency must be more than twice the highest baseband frequency. Otherwise, as shown by Fig. 7.15(b), overlap, or *aliasing*, will occur and then demodulation will not be possible. Pulse amplitude modulation suffers from the same disadvantage as AM in that its signal-to-noise ratio is not very good.

Example 7.11

Rectangular pulses of $2\mu s$ duration and pulse repetition frequency 20 kHz are amplitude modulated by a 5 kHz sinusoidal voltage and then applied to the input terminals of a low-pass filter. Determine the frequency components present at the output of the filter if it has a sharp cut-off at (*a*) 3 kHz, (*b*) 10 kHz and (*c*) 25 kHz.

Solution
The PAM waveform contains components at (i) 5 kHz, (ii) 20 ± 5 = 15 kHz and 25 kHz, (iii) 40 ± 5 = 35 kHz and 45 kHz, etc.
(*a*) No output.   (*Ans.*)
(*b*) 5 kHz.   (*Ans.*)
(*c*) 5 and 15 kHz.   (*Ans.*)

Pulse duration modulation

Pulse duration modulation (PDM) employs pulses of constant amplitude and constant spacing between their leading edges but whose width is modulated by the baseband signal. The varying pulse width is obtained by modulating the position in time of the trailing edge of each pulse. The frequency spectrum of a PDM wave also includes a baseband component, plus a number of higher frequency, unwanted, components, and this means that a PDM wave can be demodulated by passing it through a low-pass filter of appropriate cut-off frequency.

Example 7.12

A PDM system can transmit sinusoidal signals of maximum amplitude 20 V and maximum frequency 10 kHz. State (*a*) the minimum sampling frequency, (*b*) the maximum pulse duration, that can be used. (*c*) A 5 kHz, 10 V sine wave is transmitted over this system. Draw the PDM waveform.

Solution

(*a*) $f_s = 20$ kHz.   (*Ans.*)

(*b*) Maximum pulse duration $= 1/f_s = 1/(20 \times 10^3)$
$$= 50 \ \mu s.   (Ans.)$$

(*c*) Assuming that a $+20$ V signal produces a pulse duration of $50 \ \mu s$ and a -20 V signal produces pulse duration of zero, then (i) 0 V gives $25 \ \mu s$ pulse duration, (ii) $+10$ V gives $37.5 \ \mu s$ duration, and (iii) -10 V gives $12.5 \ \mu s$ duration. Using these figures the PDM wave has been drawn and it is shown in Fig. 7.17.

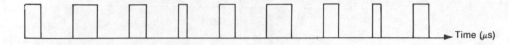

Fig. 7.17

Time (μs)

Pulse position modulation

Pulse position modulation (PPM) employs pulses of constant amplitude and width whose position in time is modulated by the baseband signal. A PDM wave is first generated and then it is differentiated to produce a series of narrow, alternately positive and negative, pulses. The positive pulses correspond with the fixed-position leading edges, and the negative pulses correspond with the variable-position trailing edges, of the PDM pulses. The negative pulses therefore form the PPM waveform and so the positive pulses must be removed. The removal of the positive pulses is achieved by applying the PPM signal to a rectifier circuit. The PPM waveform contains components in FM sidebands positioned either side of the PRF and either side of integer multiples of harmonics of the PRF. The baseband signal itself is not present and so the PPM signal cannot be demodulated by a simple low-pass filter. Instead, demodulation of a PPM signal is carried out by first passing the wave through a low-pass filter to remove all components except the FM sidebands of the PRF and then applying the output of the filter to an FM demodulator. Alternatively, the received PPM signal can be converted into the corresponding PDM signal and this can then be demodulated by a low-pass filter. The basic block diagram of a PPM system is shown by Fig. 7.18.

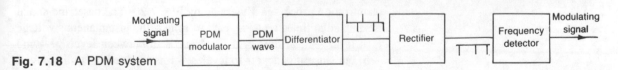

Fig. 7.18 A PDM system

Pulse code modulation

Pulse code modulation (PCM) is a digital system in which the baseband signal is sampled at regular intervals and information about each sample is transmitted using binary-coded pulses. The permissible range of baseband voltages is divided up into a number of *quantum levels*.

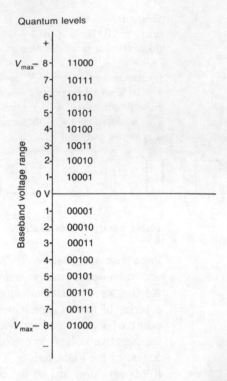

Quantum levels

+		
V_{max}	8	11000
	7	10111
	6	10110
	5	10101
	4	10100
	3	10011
	2	10010
	1	10001
0 V		
	1	00001
	2	00010
	3	00011
	4	00100
	5	00101
	6	00110
	7	00111
V_{max}	8	01000
−		

Baseband voltage range

Fig. 7.19 Quantum levels

Each quantum level is allocated a number as shown by Fig. 7.19 in which 16 levels have been drawn. Each quantum level has the binary-coded number shown. The most significant bit is used to indicate whether the sampled voltage is positive or negative; binary 1 in this position indicates a positive voltage while binary 0 indicates a negative voltage. The number of levels employed is always equal to 2^n, where n is the number of bits per digital word. Since there are 5 bits per digital word shown in Fig. 7.19 there could be as many as 32 quantum levels.

The baseband signal is sampled at regular intervals to produce a PAM signal. The amplitude of each of the PAM pulses thus obtained is then rounded off, or *quantized*, to the nearest quantum level and then the number of that level is encoded into binary form. The quantization process is illustrated by Fig. 7.20. The baseband signal is sampled at times t_1, t_2, t_3, etc. At time t_1 the instantaneous voltage is at level -3, at time t_2 the voltage is in between levels -4 and -5, but since it is nearer to level -4 it is rounded off to that level. Similarly, at time t_3 the instantaneous baseband voltage is between levels -4 and -3 but it is again rounded off to level -4. The quantization of the baseband signal is carried out at each of the times shown in Fig. 7:20 to give the values listed in Table 7.3.

The digital waveform that is sent to the line to signal these binary numbers is shown by Fig. 7.21. A pulse is transmitted to represent binary 1 and no pulse is sent to indicate binary 0. In most practical systems a more complicated code is used for the signal sent to line.

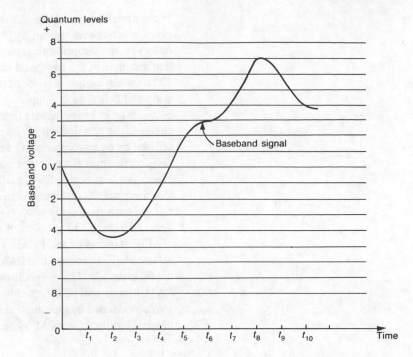

Fig. 7.20 Digital waveform

Table 7.3

Time	t_1	t_2	t_3	t_4	t_5	t_6	t_7	t_8	t_9	t_{10}
Quantum level	−3	−4	−4	−1	2	3	4	7	6	4
Binary number	00011	00100	00100	00001	10010	10011	10100	10111	10110	10100

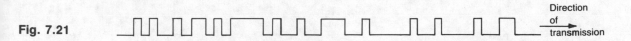

Fig. 7.21

Direction of transmission

Example 7.13

The CCITT 30 channel PCM system uses 256 quantum levels. (*a*) How many bits are used to represent each sample of the baseband signal? (*b*) Which quantum levels are represented by (i) 00001100 and (ii) 10000111?

Solution
(*a*) $256 = 2^n$, so $n = 8$. (*Ans.*)
(*b*) (i) −12. (ii) +7. (*Ans.*)

Time-division multiplex

The minimum sampling frequency employed in a PCM system must be at least twice the highest frequency contained in the baseband signal.

For commercial-quality speech the highest frequency is 3400 Hz which gives a minimum sampling frequency of 6800 Hz. In practice, however, the sampling frequency employed is 8000 Hz. This means that a sample of the baseband signal is taken every 1/8000 s or every 125 μs. Each sample is encoded using eight bits each of which is 488 ns wide and hence each sample occupies a time period of 3.9 μs. Thus only 3.9 μs in every 125 μs time interval is occupied by a particular channel, the remaining 121.1 μs is unoccupied and this space can be utilized by other channels. The number of channels that can be time-division multiplexed together is equal to 125/3.9 = 32. Two of these channels are allocated for signalling and synchronization purposes so that there are 30 channels available to carry speech signals. The bit rate on each channel is 8 $\times$ 8000 = 64 kb/s and the bit rate on the *bearer circuit* is 32 $\times$ 8 $\times$ 8000 = 2.048 Mb/s.

At the far end of the PCM system the received PCM signal must be decoded to recover the PAM signals that are proper to each of the 30 channels. If the synchronization of the system is correct the PAM signals will be directed to their correct channels and here they are demodulated by passing them through a low-pass filter. The basic block diagram of a PCM system is shown by Fig. 7.22.

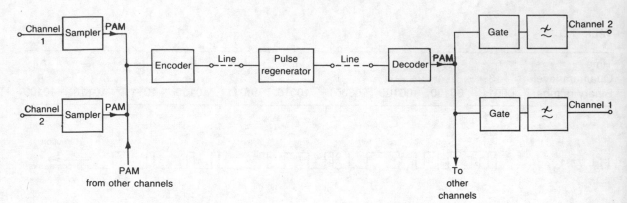

Fig. 7.22 A multi-channel PCM system

Bandwidth

The minimum bandwidth necessary to transmit a PCM system is equal to the maximum fundamental frequency of the digital signal sent to line. The maximum fundamental frequency occurs when alternate 1 and 0 pulses are transmitted and then it is equal to one-half of the bit rate.

$$\text{Bandwidth} = \text{(bit rate)}/2 \qquad (7.13)$$

Pulse regeneration

A digital signal is both attenuated and distorted as it travels along a transmission line. The PCM receiver must be able to determine accurately at each instant whether it is receiving a binary 0 or a binary 1 bit. A badly attenuated and/or distorted pulse may be mistaken as being at the logical 1 level instead of the logical 0 level, or vice versa. To overcome this potential problem the pulses travelling from one end of the system to the other are *regenerated* at regular intervals along the line. This ensures that noise and distortion are not cumulative and that the overall performance of the system is independent of the length of the system.

When data signals are to be transmitted over a PCM system they can be first converted to VF form and then handled in the same way as speech signals. Alternatively, the 64 kb/s capacity of a channel can be used to provide a high-speed data circuit that can be multiplexed to give many more lower-speed channels. The channelling equipment is not then required. The full 2.048 Mb/s capacity of a PCM bearer circuit can also be used to carry data; the high-speed data circuit is connected directly to the line and no terminal equipment is needed. The user of such a wide-band circuit usually employs multiplexers to obtain a large number of circuits.

Appendix A

The instantaneous voltage v of an AM wave is

$$v = (V_c + V_m \sin \omega_m t) \sin \omega_c t \qquad (7.1)$$

$$= V_c \sin \omega_c t + V_m \sin \omega_c t \sin \omega_m t$$

A trigonometric identity is

$$2 \sin A \sin B = \cos (A - B) - \cos (A + B)$$

Let $A = \omega_c t$ and $B = \omega_m t$ then

$$v = V_c \sin \omega_c t + (V_m/2) \cos (\omega_c - \omega_m)t$$
$$- (V_m/2) \cos (\omega_c + \omega_m)t \qquad (7.14)$$

8 Control systems

A control system is employed to regulate, or govern, an output quantity. The output quantity might be the electrical, mechanical, chemical, etc. energy, supplied to a load, or the control of a process. Essentially, a control system is an arrangement of a number of various items of equipment that is employed to control either itself or some other system. The size of a control system may vary enormously, ranging from, say, an electric toaster to a large manufacturing plant. Each of the parts of a control system may simply consist of a single device, such as a temperature sensor, or a circuit, such as an amplifier, or it may itself be a smaller control system. Examples of control systems are many and varied and include motor speed control, the remote positioning of a shaft, of a door, of an aerial, etc., voltage regulation, temperature control, liquid level and/or pressure control, the control of machine tools and so on.

Before the output quantity can be controlled by the system it must first be accurately measured and the measured value converted into an electrical signal, some kind of sensor often being required at the output. The sensor will be some form of *transducer*, which is the name given to any device that is able to convert the output quantity of a control system into the equivalent electrical signal. The electrical output of the transducer is fed back to the input of the system and here it is subtracted from an input electrical signal which represents the wanted output quantity. The input signal is used to set the required value of the output quantity. The difference between the two electrical signals is known as the *error signal* and it is used to operate the system in such a way that the error in the output quantity is reduced.

Open-loop and closed-loop systems

Two basic types of control system are possible and they are known as the *open-loop system* and the *closed-loop system*. An open-loop system is one in which the output of the system does not depend in any way upon the value of the output quantity. The block diagram of such a system is shown by Fig. 8.1. A commonly met example of an open-loop system is an electric light in the home. When the light is switched ON it will remain on, regardless of the light level in the room, until such time as it is switched OFF. The light is not switched off automatically if the light level in the room becomes high

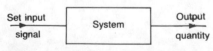

Fig. 8.1 Open-loop control system

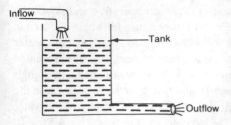

Fig. 8.2 Open-loop water tank system

enough for the illumination to be no longer needed. Another example of an open-loop system is an electric fire; once it has been switched ON it will remain on regardless of the room temperature until someone turns it OFF. The open-loop system is employed for these, and similar purposes, because it is simple, cheap and there is no risk of instability, i.e. the light will not get continually brighter or the fire get continually hotter as time goes on. Two further examples of open-loop systems are shown in Figs 8.2 and 8.3. Figure 8.2 shows an open-loop water tank system. Water flows into the tank via an inflow pipe mounted above the tank, and it flows out of the tank via another pipe at the bottom of the tank. If the rate of water inflow is constant the water level in the tank will rise until the outflow is equal to the inflow. If, then, the inflow is increased the water level will rise still further and if the inflow is reduced the water level will fall. There is, however, no automatic control of the water level at any desired height.

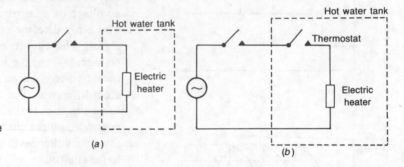

Fig. 8.3 (a) Open-loop temperature control system; (b) closed-loop temperature control system

Figure 8.3(a) shows a hot water tank whose water temperature is controlled by turning a heater switch ON and OFF. Clearly there is no automatic control of the water temperature. The effectiveness of the system could be greatly improved by the addition of a thermostat switch as shown by Fig. 8.3(b). This is now a simple form of closed-loop system. The temperature of the water inside the tank is monitored by the thermostat and this is set to open when the water temperature reaches the desired value and so disconnect the power. When the heater is first switched ON the water temperature will be less than the setting of the thermostat and so the thermostat contacts will be closed. Current will flow in the heater and the water will be heated. When the temperature of the water reaches the set value the thermostat contacts will open to disconnect the power supply so that no current

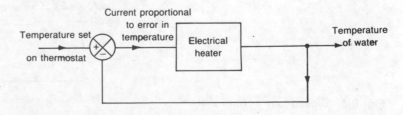

Fig. 8.4 Block diagram of a closed-loop temperature control system

flows in the heater. The heater is then non-operative and hence the water cools down. When the water temperature falls below a certain value the thermostat contacts will again close to switch ON the heater and so on. The closed-loop hot water tank system can be represented by the block diagram shown in Fig. 8.4. It can be seen that only signals appear in a block diagram, power inputs are not shown. Because the heater can only be turned either on or off the temperature of the water will vary either side of the desired, set, value (see Fig. 8.5).

Closed-loop control systems

The generalized block diagram of a closed-loop control system is given by Fig. 8.6. The functions of the blocks are as follows:

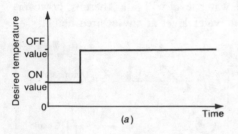

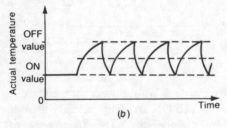

Fig. 8.5 Action of an OFF/ON closed-loop temperature control system

(a) The *measurement block* converts information about the output quantity into the corresponding electrical signal. This signal may be in either analogue or digital form and it may be either a voltage or a current.

(b) The error detector compares the values of the electrical signals which represent the desired output quantity — the set signal — and the actual output quantity and produces an output voltage whose magnitude is directly proportional to their difference. The difference output voltage is generally known as the error signal. If the error is positive, i.e. the controlled output quantity is greater than the set input value, the error voltage will have the polarity which, when applied to the controller, will tend to reduce the error. Conversely, if the error is negative the output voltage of the error detector will have the opposite polarity. In a computer-controlled system the generation of the error signal is performed by software.

(c) The *controller* is actuated by the error signal and it determines the changes in the controlled system that are necessary to correct the error in the output quantity. The output signal of the controller drives the control device to make the output quantity vary in the direction that will reduce the error. The controller may be any device or circuit that is able to vary the power supplied to a load under the control of an electrical

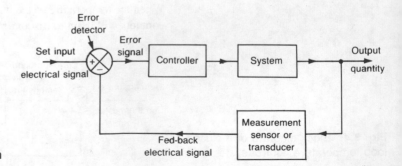

Fig. 8.6 Closed-loop control system

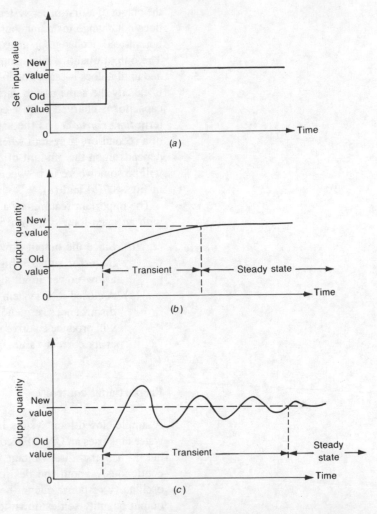

Fig. 8.7 (*a*) Step function input; (*b*) output of a first-order system; (*c*) output of a closed-loop system

input signal. The output power may vary continuously in direct proportion to the input signal, or it may vary in discrete steps (perhaps just on and off like the thermostatically controlled hot water tank in Fig. 8.3(*b*)). The controller may be a hydraulic or a pneumatic device, but most often it is some form of electronic amplifier or other electrical device. Commonly, a small input voltage is used to produce a large output current which is used to control an electric heater or an electric motor.

The performance of a closed-loop system is tested by making a sudden change to the set input value. Such a change, known as a *step function*, will produce an output response that depends upon whether the system is of the *first order* or of the *second order*. A first-order system is one that contains only one device that is able to store energy, a second-order system has two or more such devices. Figure 8.7(*a*) shows an input step function and Fig. 8.7(*b*) shows the response at the output of a first-order system. It can be seen that the output quantity does not change instantaneously from its old value to the new value but, instead, it changes in an exponential manner from one to the other. The output quantity takes some time to reach its *steady-state value* and until it does is said to be in the transient state. The output response is exactly the same as the variation of the capacitor voltage when a capacitor is charged from a constant-voltage source (p. 65) and the term *time constant* has the same meaning. The variation with time of a second-order system when a step function is applied to its input depends upon the amount of *damping* in the system (p. 180) and it will be somewhere in between, and including, the responses shown in Figs 8.7(*b*) and (*c*).

The important features of a closed-loop control system compared with an open-loop system are:

(*a*) Since the output power is controlled, but not supplied, by the error signal a high-power output can be controlled by a low-power input signal.

(*b*) A closed-loop system is self-regulating. This means that any disturbances, e.g. noise, to the system, and/or to the load will produce an error signal that will keep the output quantity at its correct value.

Proportional control

A simple closed-loop system like the thermostatically controlled hot water tank uses an ON/OFF controller. This is simple in its operation but it has the disadvantage of giving an output that oscillates continuously about the desired value. For most systems such an oscillatory response cannot be tolerated since it is required that the output quantity settles down to its steady-state value within a fairly short period of time. This means that a controller must be employed that produces an output that is directly proportional to its input (error) signal. The system is known as *proportional control*. Some examples of proportional control systems are given in Figs. 8.8–8.17.

Water level control

Figure 8.8 shows a water tank that is supplied with water pumped in by an electric pump through a pipe mounted above the tank. Water leaves the tank via an outflow pipe at the bottom of the tank. The water level inside the tank is to be maintained constant as the rate of water outflow is varied. When the water level is correct, i.e. at its set value, the current that flows in the control winding of the electric motor is set to zero by adjustment of the reference voltage V_{REF}. The electric motor, and hence the pump, will then run at the speed dictated by the current that flows in the main field winding alone. When water is taken from the tank the water level in the tank falls and this results

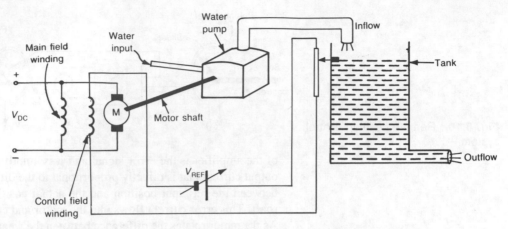

Fig. 8.8 Water level control system

in a current flowing through the control winding. The direction of this current is such that the magnetic field it sets up around the control winding weakens the main magnetic field and this causes the motor to increase its speed. The pump then operates at faster speed and delivers an increased flow of water into the tank. When the water level rises above the set value a current flows in the control winding in the opposite direction to before and the magnetic field that this current produces aids the main magnetic field; this makes the motor, and hence the pump, slow down to reduce the water inflow to the tank. The block diagram of the system is shown in Fig. 8.9.

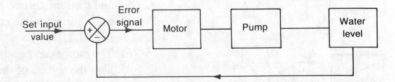

Fig. 8.9 Block diagram of the
system shown in Fig. 8.8

Remote position control

A remote position control (RPC) system is employed to control accurately the position θ_o of a shaft, a radar aerial or some other large mechanical body. The basic arrangement of an RPC system is shown by Fig. 8.10. The desired position of the output shaft is set by moving the input control to an angular position θ_i. This angular position is converted into a voltage by means of a slider that is mounted on the input control and that moves around the track of the input potentiometer. For any position θ_i of the input control a voltage, in the range $\pm V$ volts, is applied to one of the input terminals of the amplifier. The position θ_o of the output shaft is monitored by a slider on the output potentiometer and it is represented by another voltage, also in the range $\pm V$ volts. This voltage is fed back to the input of the system to be applied to the other input terminal of the amplifier. The difference between the two voltages applied to the input terminals

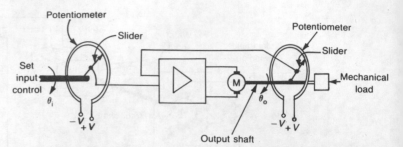

Fig. 8.10 Remote position control system

of the amplifier is the error signal and it is amplified to produce an output current that is directly proportional to the difference, $\theta_i - \theta_2$, between the set input position and the actual position of the output shaft. This error current flows into the motor and causes it to rotate. As the motor rotates the difference between the wanted and the actual positions of the output shaft becomes smaller. Eventually the error is zero, and then there is zero error signal, zero current is supplied to the motor and then the motor will stop. There is always a small range of errors over which the system cannot operate because the motor torque that is produced is too small to overcome the frictional forces that exist within the system. This range is known as the *dead zone*.

The motor can be controlled in either one of two ways since the torque T developed by a d.c. motor is equal to $T = KI_fI_a$ where K is a constant, I_f is the field current and I_a is the armature current.

(a) The motor can be supplied with a constant armature current and then the torque produced by the motor will be directly proportional to the current flowing in the field winding (see Fig. 8.11(a)).

(b) The motor can be supplied with a constant field current and then the torque of the motor is determined by the armature current (see Fig. 8.11(b)). In either case the amplifier must be able to supply the full power required by the motor field, or armature, and field control is generally employed for powers of up to about 1500 W and armature control for higher-powered motors.

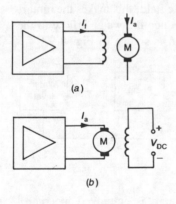

Fig. 8.11 Control of a motor: (a) constant armature current, variable field current; (b) constant field current, variable armature current

The output of an RPC system will have zero error when the set input is varied and the steady-state condition has been arrived at, but zero error will not be achieved if the input control is rotated continuously. If, however, the load applied to the output shaft is suddenly changed to alter the output position θ_o the system will operate to reduce the error but it will not be able to reduce it to zero; a small error will then exist that can only be eliminated by an alteration in the set input value θ_i.

Very often a specially designed motor, known as *servomotor*, is used for RPC applications. The servomotor differs from a conventional d.c. motor in that its armature is made both longer and narrower than is usual and this gives it a high torque along with low inertia.

Motor speed control

The speed of a d.c. series motor can be controlled by varying the d.c. voltage applied to it. The speed of a shunt d.c. motor can be controlled by either (a) applying a constant voltage to the field winding and then varying the armature current, or (b) supplying the armature with a constant current and then varying the current passed through the field winding. Figure 8.12 shows the elements of a motor speed control system. The speed of rotation, ω_o radians/second, of the output shaft is converted into a proportional voltage by means of a *tachogenerator* which is mounted on the shaft. The tachogenerator is a simple d.c. or a.c. generator which generates an output voltage that is directly proportional to the velocity of the output shaft. The tachogenerator voltage is fed back to the input of the system and here it is applied to one of the inputs of the amplifier. A d.c. reference voltage V_{REF} is applied to the input terminal of the other amplifier. The difference between these two voltages, which is the error signal, is amplified by the amplifier to produce an output current that is passed through the field winding of the motor to control the motor's speed. There must always be a small error between the actual speed of the motor and the set input speed in order to produce a controlling field current.

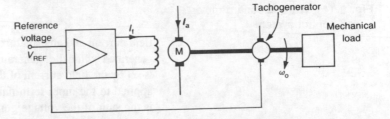

Fig. 8.12 Motor speed control system

Another motor speed control system, suitable for the control of large power motors, is shown by Fig. 8.13. A six-thyristor rectifier circuit is used to rectify a three-phase supply to provide the large d.c. current needed by the armature of the motor. The field current is supplied by a rectified single-phase supply. Once the thyristors in the circuit have been turned on the mean output voltage, and hence the armature current of the motor, can be controlled by varying the conduction angle of the thyristors.† The conduction angle is selected by the control circuit as it responds to the error voltage output of the voltage comparator.

A commonly-employed motor speed control system is the Ward—Leonard system shown by Fig. 8.14. A d.c. generator is driven at a constant speed by an induction motor so that its generated e.m.f. is directly proportional to the current flowing in its field winding. The motor whose speed is to be controlled is supplied with a constant

† Electronics III.

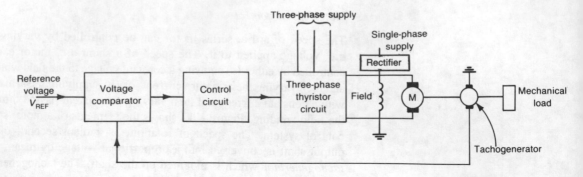

Fig. 8.13 Speed control of a large-power motor

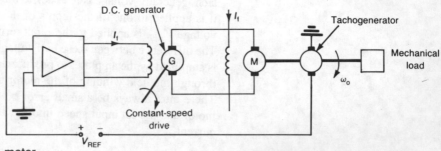

Fig. 8.14 Ward–Leonard motor-speed control system

field current and an armature current that is determined by the e.m.f. generated by the d.c. generator. This means that the speed of the motor is set by the field current of the d.c. generator and hence by the signal applied to the input terminals of the amplifier. This signal, in turn, is the sum of the voltage generated by a tachogenerator mounted on the shaft of the motor and a reference voltage V_{REF}. This system of motor speed control allows power outputs of several kilowatts to be controlled by just a few milliwatts and it is suitable for applications which involve high output powers over a range of different speeds.

Voltage regulation system

Figure 8.15 shows how the terminal voltage of a d.c. generator can be maintained within close limits as the load on the generator is varied. The terminal voltage, or a fraction of it, is fed back to the input of the system and here it is compared with a reference voltage V_{REF}. The voltage difference between the two voltages, the error signal,

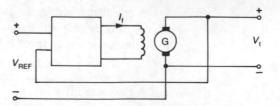

Fig. 8.15 Voltage regulation of a d.c. generator

is then amplified to produce the field current for the d.c. generator. When there is an increase in the current taken from the output terminals of the generator, because of a change in the load, there will be an increased voltage drop across the armature resistance of the generator so that the terminal voltage will fall. The difference between the fed-back voltage and the reference voltage will then be increased and the amplified error signal will be larger, leading to an increase in the field current. The voltage generated by the d.c. generator will then be increased and this increase will tend to keep the terminal voltage at a constant value.

Systems with several controls

Many industrial control systems have several output quantities that must be controlled and Figs 8.16 and 8.17 give two examples. Figure 8.16 shows a process system for the making of biscuits. Suitable amounts of the biscuit mixture are deposited, in the desired shape, on to a moving belt which transports the biscuits through an oven and thence to a container. Three controls are required by this system: a feed control to deposit the correct amounts of biscuit mixture at the correct instants in time on to the moving belt, a speed control to ensure that the biscuits are in the oven for the time required for their correct cooking; and a temperature control to keep the oven at the necessary temperature. Sensors are used to detect the speed of the moving belt and the temperature of the oven and these sensors feed their data to the computer. The software in the computer processes the information and then issues appropriate commands over the control lines to keep the system operating correctly. If different kinds of biscuit

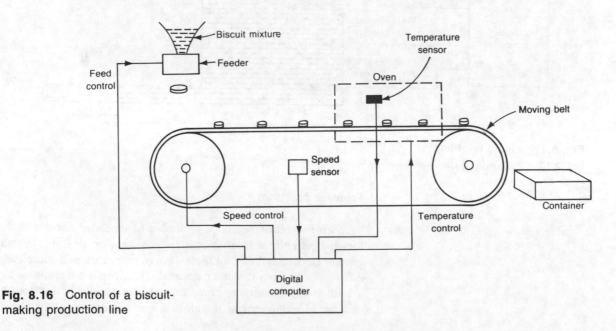

Fig. 8.16 Control of a biscuit-making production line

are made at different times the operation of the system can easily be altered as required by the use of another computer program. If any of the sensors are analogue types (most likely), their analogue output voltage will need to be converted into digital form before it is supplied to the computer; this is the function of an analogue-to-digital converter. Also, the digital commands of the computer will have to be converted from digital to analogue format before being fed to the control; this is the function of a digital-to-analogue converter.

The control system shown by Fig. 8.17 consists of a large tank that contains a liquid that must be kept at a particular level and temperature. Sensors are used to detect both the level and the temperature of the liquid in the tank and these feed information to the computer. The rate of the outflow from the tank can be varied by a valve which is controlled by the computer. If the level of the liquid in the tank becomes too low the computer increases the speed of the electric pump and more liquid is pumped into the tank. The temperature of the liquid is continually monitored by the computer and using the data thus gained the computer will vary the current that flows in the heater to maintain the temperature at the required value.

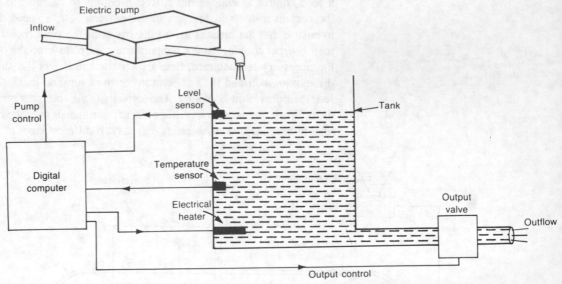

Fig. 8.17 Liquid level and temperature control

Transfer functions

A control system consists of a number of cascaded devices in the forward path plus a feedback path which may, or may not, contain another device, and each of these devices may work in a completely different way from the other devices. The overall operation of the system is therefore quite complex and it is usually easier both to understand and to analyse if a block diagram is drawn in which each

Fig. 8.18 Transfer function; sinusoidal input

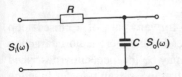

Fig. 8.19

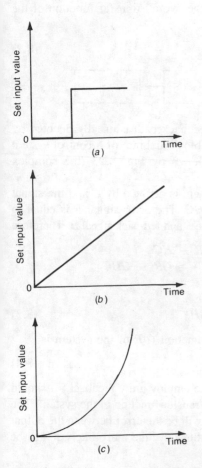

(a)

(b)

(c)

Fig. 8.20 Non-sinusoidal input signals: (a) step function; (b) ramp function; (c) parabolic function

Fig. 8.21 Transfer function; non-sinusoidal input

of the devices is represented by its *transfer function*. The transfer function of a device indicates how that device will perform when it is supplied with different input signals and it is the mathematical expression that relates the output quantity of the device to its input quantity. The transfer function of an amplifier will be expressed in units of volts/volts or amperes/volts, of an electric motor in r.p.m./ampere, of a tachogenerator in volts/r.p.m. and so on. If the input signal is sinusoidal it can, as in Fig. 8.18, be represented by $S_i(\omega)$. If the output signal is then $S_o(\omega)$ the transfer function will be

$$G(\omega) = S_o(\omega)/S_i(\omega) \tag{8.1}$$

Example 8.1

Determine the transfer function of the *RC* network given in Fig. 8.19.

Solution
$$V_o(\omega) = [V_1(\omega)(1/\omega C)]/\sqrt{[R^2 + (1/\omega C)^2]} = V_1(\omega)/\sqrt{[1 + \omega^2 C^2 R^2]}$$
$$G(\omega) = V_o(\omega)/V_1(\omega) = 1/\sqrt{[1 + \omega^2 C^2 R^2]} \quad (Ans.)$$

Input signals

In many cases the input signal to a control system is not of sinusoidal waveshape and then the concept of complex frequency must be used. The other more commonly occurring input signal waveforms are shown by Fig. 8.20. Figure 8.20(a) shows a step function. This signal is a sudden change, an increases or a decrease, in the value of the set input. In practice, the step function must always have a small rise time (p. 76), but usually this is ignored. A *ramp input* is shown in figure 8.20(b); it has a magnitude that increases or decreases linearly with increase in time. The main example of a ramp function is the rotation, at a constant velocity, of the input control of an RPC system. Less common is the input signal shown by Fig. 8.20(c). This is a parabolic input signal which occurs when the input control of a system is rotated with a constant acceleration.

The transfer function of a device for a non-sinusoidal input signal can readily be obtained if the concept of *complex frequency S* is used. Using complex frequency the complex impedance of a pure inductor can be written as SL, of a practical inductor as $R + SL$ and of a capacitor can be written as $1/SC$. Then (Fig. 8.21) the transfer function $G(S) = S_o(S)/S_i(S)$.

Example 8.2

Determine the transfer function of the circuit shown in Fig. 8.19 using complex frequency.

Solution

$V_0(S) = V_1(S)(1/SC)/(R + 1/SC) = V_1(S)/(1 + SC)$

$G(S) = V_0(S)/V_1(S) = 1/(1 + SCR)$ (*Ans.*)

When the various devices in the forward path of a closed loop control system are connected in cascade, and there is no interaction between them, each of the devices may be independently described by its transfer function. This means that the overall transfer function of the forward path is given by the product of the individual transfer functions of each device. Thus the overall transfer function of the forward path shown in Fig. 8.22 is

$$G(S) = K_1 K_2/(1 + SCR) \qquad (8.2)$$

Fig. 8.22

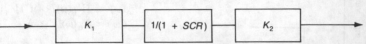

The transfer function of the feedback path is usually denoted by $H(S)$ and hence the generalized block diagram of a control system using transfer functions is as shown by Fig. 8.23 (the complex frequency S has been omitted).

The output quantity of the system is denoted by C and the signal B fed back to the input is $B = HC$. The error signal E is equal to the difference between the set value R and fed-back signal B. Therefore

$$E \qquad = R - B = R - HC$$
$$C = GE \qquad = G(R - HC) = GR - GHC$$
$$C(1 + GH) = GR$$

and

$$C \qquad = GR/(1 + GH) \qquad (8.3)$$

Therefore, the overall transfer function C/R of the system is

$$C/R = G/(1 + GH) \qquad (8.4)$$

When $H = 1$ the system is said to employ unity feedback. It should be noted that G is the open-loop transfer function of the system. The larger the value of G the smaller will be the error between the output C and the set input R but, unfortunately, the more likely it will be that the system will be unstable.

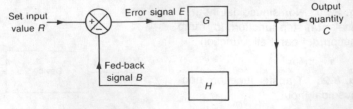

Fig. 8.23 Block diagram of a closed-loop system using transfer functions

Example 8.3

The motor speed control system shown in Fig. 8.13 has the following transfer functions: amplifier, 12 V/V; motor, 120 r.p.m./V; tachogenerator, 0.08 V/r.p.m. If the reference voltage is 80 V calculate the steady-state speed of the motor.

Solution

Working from first principles: if the speed of the motor is N r.p.m. then tachogenerator voltage $= 0.08\,N$ V.

Error voltage is $80 - 0.08\,N$ V.

Amplifier output voltage $= 12(80 - 0.08N) = 960 - 0.96\,N$ V.

Motor speed $N = 120(960 - 0.96N) = 115.2 \times 10^3 - 115.2N$ r.p.m.

Therefore, $116.2N = 115.2 \times 10^3$, or

$\qquad N = 991.4$ r.p.m.　　(*Ans.*)

Alternatively, using equation (8.3),

$\qquad G = 12$ V/V $\times$ 120 r.p.m./V $= 1440$ r.p.m./V.

$\qquad H = 0.08$ V/r.p.m.

Therefore, $C = (1440 \times 80)/[1 + (1440 \times 0.08)]$

$\qquad\qquad = 991.4$ r.p.m.　　(*Ans.*)

Example 8.4

The voltage regulation system shown in Fig. 8.15 has the following transfer functions: amplifier, 0.2 A/V and generator 50 V/A. The armature resistance is 0.1 Ω. The terminal voltage of the generator is to be maintained to within the limits 100 ± 1 V as the current taken by the load varies from zero to 200 A. Calculate the necessary value of the d.c. reference voltage.

Solution

The terminal voltage V_t is

$\qquad V_t = 50I_f - I_L\,r_a$　　　　　　　　　　　　　　　(8.5)

$\qquad I_f = 0.2(V_{REF} - V_t)$　　　　　　　　　　　　(8.6)

On full load the voltage dropped across the armature resistance is $200 \times 0.1 = 20$ V and so the increase in the generated voltage that is required is $99 - 80 = 19$ V. Therefore, from equation (8.5),

$\qquad 99 = 50I_f - 20$, $I_f = 119/50 = 2.38$ A.

Substituting this value of the field current into equation (8.6) gives

$\qquad 2.38 = 0.2(V_{REF} - 99) = 0.2V_{REF} - 19.8$.

Hence $V_{REF} = 22.18/0.2 = 110.9 \simeq 111$ V.　　(*Ans.*)

On no-load $V_g = V_t = 101$ V so $101 = 50I_f$ and $I_f = 101/50 = 2.02$ A.

Hence $2.02 = 0.2(V_{REF} - 101) = 0.2V_{REF} - 20.2$ and

$\qquad V_{REF} = 22.22/0.2 = 111.1 \simeq 111$ V.　　(*Ans.*)

Alternatively, using equation (8.3),

$G = 0.2$ A/V $\times$ 50 V/A $= 10$ and $H = 1$.

Therefore, $101 = 10V_{REF}/(1 + 10) = (10/11)V_{REF}$ and

$\qquad V_{REF} = 111.1 \simeq 111$ V.　　(*Ans.*)

Proportional plus integral control

The steady-state error in the output quantity of a proportional control system can be eliminated by the use of *integral control* in addition

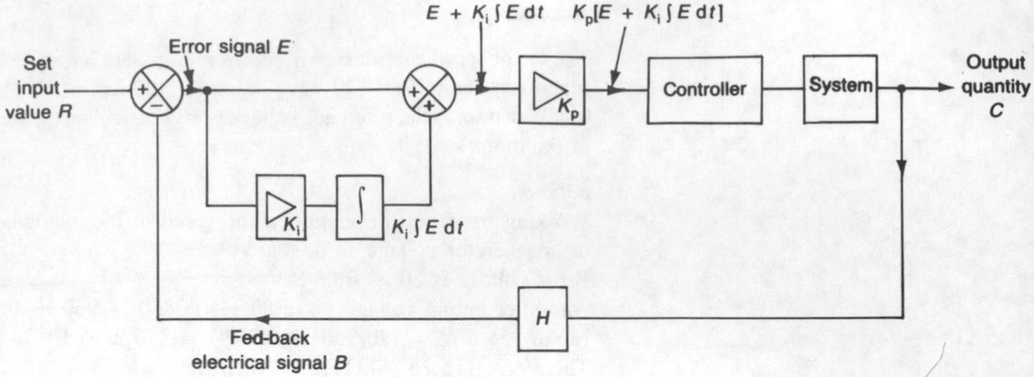

$$E + K_i \int E \, dt \qquad K_p[E + K_i \int E \, dt]$$

Error signal E

Set input value R

K_p

Controller

System

Output quantity C

K_i

$\int$

$K_i \int E \, dt$

H

Fed-back electrical signal B

Fig. 8.24 Proportional plus integral control

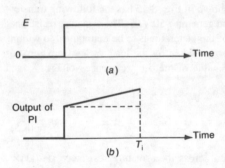

E

0 → Time

(a)

Output of PI

→ Time

T_i

(b)

Fig. 8.25 Operation of a PI system

to proportional control. The block diagram of a proportional plus integral (PI) control system is shown in Fig. 8.24. The signal applied to the controller is now equal to the sum of a proportional signal and a signal that is proportional to the integral, with respect to time, of the error. Thus the total signal applied to the controller is

$$K_p[E + \int K_i E \, dt] \qquad (8.7)$$

where K_p is the gain of the proportional circuit and $K_i = 1/T_i$, where T_i is known as the *integral time*. The proportional control circuit produces an error signal that is directly proportional to the actual error at each instant in time. The integral control circuit gives an error signal that is the sum of the past errors. If the set signal is a step function the proportional control circuit will give a steady-state error and the integral signal will have a ramp waveform. The two waveforms are added as shown by Fig. 8.25 to reduce the offset error, the error being reduced to very nearly zero in the integral time. A positive error gives an increase in the output of the integral control circuit while a negative error leads to a decrease in the output. The main disadvantage of integral control is its tendency to make a system more unstable.

Proportional plus integral plus derivative control

The PI control system is unable to deal effectively with a rapid change in the output quantity of the system. This difficulty arises for two reasons: (*a*) the proportional control circuit cannot produce an error signal until an error of some size exists, and (*b*) the integral control circuit is slow to respond to any change at the output because its operation is based upon past errors. To enable a closed-loop system to respond quickly to changes in the output quantity *derivative control* must also be employed.

Derivative control produces a corrective signal that is proportional to the rate of change of the error signal, i.e. $C = K_d dE/dt$, where K_d is the *derivative control gain*. The corrective output signal is varied in a way that follows the rate at which the error signal is

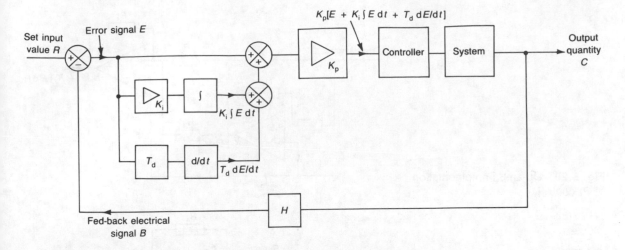

Fig. 8.26 Proportional plus integral plus derivative control

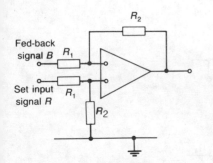

Fig. 8.27 Op-amp implementation of P control

changing and this allows the error to be quickly corrected before it becomes large. Derivative control tends to make a system more stable, and when it is used in conjunction with proportional control it allows the proportional control output to be higher than would otherwise be possible without instability occurring.

The basic block diagram of a proportional plus integral plus derivative (PID) control system, also known as *three-term control*, is shown in Fig. 8.26. The derivative control circuit produces an error signal that is directly proportional to the rate at which the error is changing. The *derivative time* T_d is the time that elapses before the derivative signal becomes equal to the proportional signal. The signal applied to the controller is

$$K_p[E + K_i \int E \, dt + T_d dE/dt] \tag{8.8}$$

The three control strategies, P, PI and PID, can be implemented using either analogue or digital techniques with a digital computer or a microprocessor. Figures 8.27–8.29 show how op-amps can be employed to implement the three control laws. Figure 8.27 shows an op-amp that has been connected as a proportional control circuit. The set value and fed-back signals are each fed via an equal-value resistor R_1 to a different input terminal of the op-amp. The difference between the two signals is therefore amplified with a gain of $-R_2/R_1$ to give an output voltage that is the required error signal. To obtain a PI control circuit the error signal produced by the circuit of Fig. 8.27 must be fed into an op-amp that has been connected to act as an integrator. The required circuit is shown in Fig. 8.28. The gain of the first op-amp is often set to unity by the use of equal-value resistors, i.e. $R_1 = R_2$; the gain of the second op-amp is $K_p = -R_4\frac{1}{2}/R_3$ and its integral time is $T_i = C_1 R_2$ seconds. Two ways in which a PID control circuit can be obtained are shown by Figs 8.29(*a*) and (*b*). A modified version of Fig. 8.29(*b*) could, of course, be used instead of Fig. 8.28.

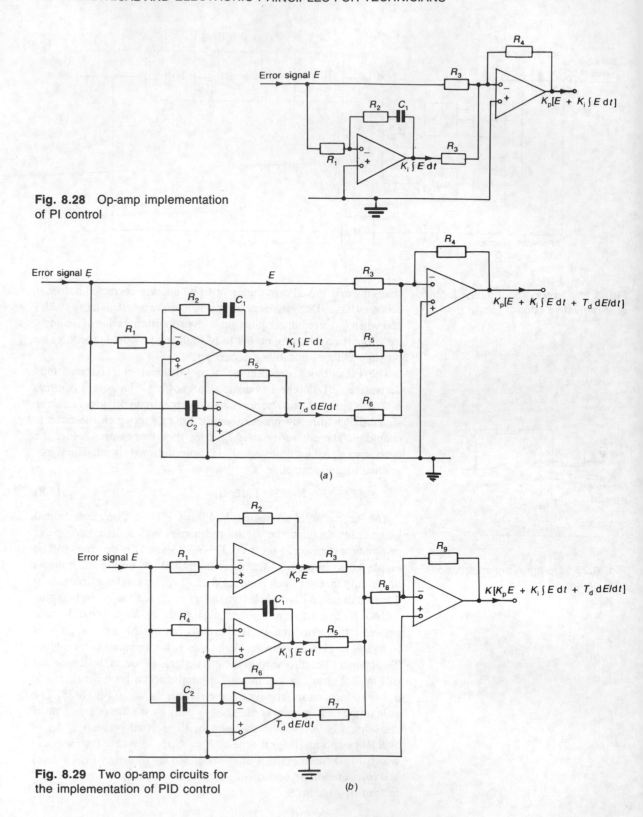

Fig. 8.28 Op-amp implementation of PI control

Fig. 8.29 Two op-amp circuits for the implementation of PID control

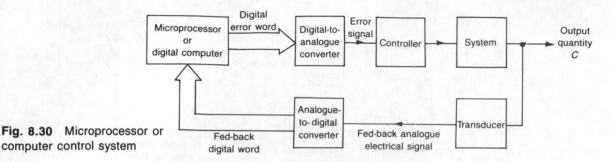

Fig. 8.30 Microprocessor or computer control system

Computer or microprocessor control

The operation of a modern closed-loop control system may often be given to a computer or a microprocessor, with software being used to implement the required one of the three control strategies. The implementation of a computer (or microprocessor)-controlled system means that all the signals to the computer must be digital in nature. Since most sensors are analogue types this means that an analogue-to-digital converter (ADC) will be required to convert the analogue signals into digital form. In the other direction, computer to controller, the control signals originated by the computer must be converted from digital into analogue form and hence a digital-to analogue converter (DAC) is also necessary.

Figure 8.30 shows the generalized block diagram of a computer-controlled system. The value of the output quantity of the system is measured by a sensor and the electrical signal thus produced is applied to an ADC. Samples of the waveform of this electrical signal are then converted into the corresponding digital word. This digital word is then passed to the computer where the software processes the information. The computer calculates the values of the required actuating signals from the values of the set input, held in memory, and the controlled output. The set input value and the actual controlled output value are each represented by a set of equations known as *algorithms*, from which the computer is able to calculate and generate the necessary digital error signal. This digital word is passed to the DAC and this converts the word into the equivalent analogue signal. This analogue signal is then applied to the controller and causes it to vary the output quantity in the direction that will reduce the error.

Figure 8.31 shows how a microprocessor may be used to control (*a*) the speed of a motor and (*b*) the temperature of an oven. The controlled speed of the motor is monitored by a tachogenerator to obtain an analogue voltage that is proportional to the output speed. This analogue voltage is applied to the ADC to be converted into the corresponding digital word. This digital word is then fed into the computer and here it is compared with a reference digital word that represents the required speed of the motor and is held in the software. Another digital word, representing the difference between the sample word and the reference word, is generated by the computer and this word is sent to the DAC for conversion into analogue form. The

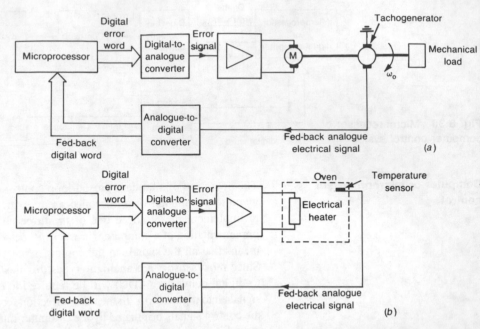

Fig. 8.31 (a) Microprocessor control of the speed of a motor; (b) microprocessor control of oven temperature

analogue output voltage of the ADC is then applied to the amplifier to be amplified to give the required motor field current.

In the case of the temperature control system the memory of the microprocessor will store the reference temperature T_{REF} °C plus a required tolerance of $\pm T$ °C. Both T_{REF} and T will be programmable and alterable as required. If the sampled temperature of the oven is higher than $(T_{REF} + T)$ °C the oven heater is to be turned ON, and if the sampled temperature is less than $(T_{REF} - T)$ °C the heater is to be turned OFF. When the ADC is ready with its data the microprocessor will execute an instruction that reads the data of the ADC into the accumulator. This data is then compared with T_{REF} and if the two temperatures differ by more than T °C a command will be sent to the DAC to turn the heater ON, or OFF, as necessary. A simple flow chart for the system, assuming that $T = 0$, is shown in Fig. 8.32. Suppose that the digital words sent to the DAC to (a) turn the oven heater ON is 01 and (b) turn it OFF is 10, that a 6800

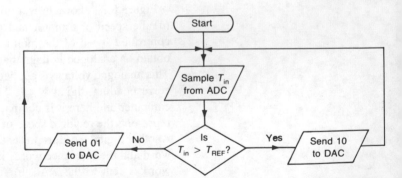

Fig. 8.32 Flow chart of the microprocessor-controlled temperature system

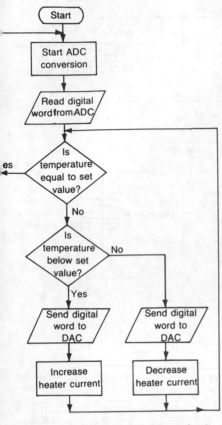

Fig. 8.33 More detailed flow chart

family microprocessor is used, the interface between the microprocessor and the ADC/DAC is via a 6821 programmable interface adaptor (PIA), and that the addresses of the input and output ports are 7001 and 7003 respectively. A simple program to control the system is:

START	0030	LDA A	7001	;input T_{in} to accumulator A.
	0033	CMP A	T_{REF}	;compare the contents of the accumulator A with T_{REF}
	0035	BGE	ABOVE	;go to ABOVE if $T_{in} > T_{REF}$.
	0037	BLT	BELOW	;go to BELOW if $T_{in} < T_{REF}$.
ABOVE	0039	LDA B	10	;load accumulator with 10.
	003B	STA B	7003	;store contents of accumulator B in PIA output.
	003E	BRA	START	;return to 0030 for next sample.
BELOW	0040	LDA B	01	;load accumulator B with 01.
	0042	STA B	7003	;store contents of accumulator B in PIA output.
	0044	BRA	START	;return to 0030 for next sample.

This simple program takes no account of the possibility that $T_{in} = T_{REF}$, nor does it allow for any tolerance centred on T_{REF}.

A more detailed flow chart for the oven heater system is given in Fig. 8.33.

The ADC operates on samples taken of the sensor's output voltage. In between the samples the feedback path is open and this means that the system is then open-loop. The sampling frequency must be at least twice the highest frequency contained in the sensor's output voltage, but it is usually chosen to be somewhat higher than that, since the higher the sampling frequency the smaller will be the overshoot of the output quantity. Because a computer or a microprocessor operates at a much faster speed than any control system the time period between successive samples can be used for the computer to control other systems. A digital computer is therefore able to act as the digital controller for several closed-loop systems at the same time, and the generalized block diagram of the arrangement is shown in Fig. 8.34. The analogue signals obtained from various sensors that are monitoring the outputs of several controlled systems are applied to a time-division multiplexer in a *data acquisition system* (DAS). On a command from the digital computer the multiplexed signal is converted into a digital word and passed on to the computer. The data word is processed by the computer to produce output data, representing the corrective actions to be taken by each of the controlled systems, that is passed to the *data output system* (DOS). The output data is converted back to analogue form and the composite analogue signal is then de-multiplexed to obtain the individual error signals. These error signals are then applied their respective controllers to reduce the error in the output quantity of each system. An example of a system in which a digital computer controls a number of variables is shown in Fig. 8.35.

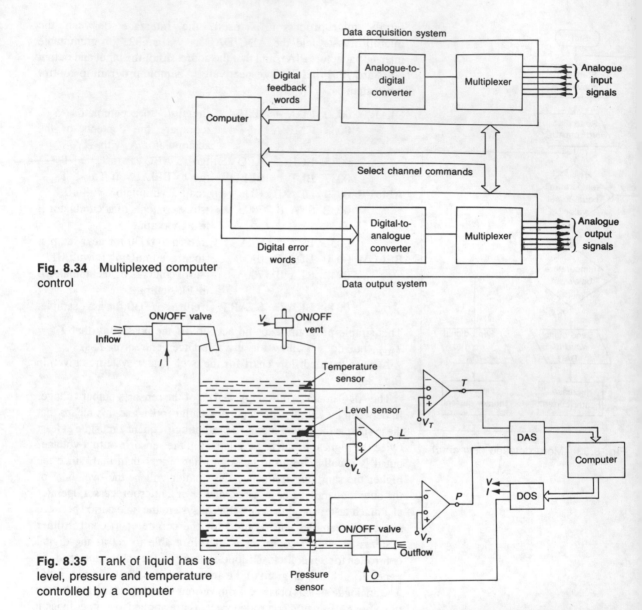

Fig. 8.34 Multiplexed computer control

Fig. 8.35 Tank of liquid has its level, pressure and temperature controlled by a computer

The computer is able to control the inflow and outflow of a liquid into and out of a tank, and maintain the liquid in the tank at any required level and at any required temperature. If the pressure in the tank exceeds a predetermined value the computer will open a vent to release some of the gas above the liquid and so lower the pressure.

Digital proportional control

The digital error is the difference between the digital word generated by the ADC and the digital word that represents the set input value

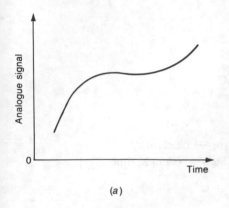

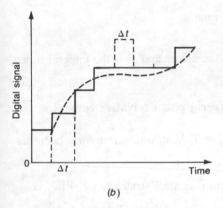

Fig. 8.36 (a) Analogue sensor signal; (b) digital representation by a DAC

held in the software. If the digital error is negative it will be in two's complement form. The range of the physical (analogue) variable that the computer can deal with is equal to the maximum count of the digital word. For an eight-bit microprocessor the range will be from 00H to FFH.

Example 8.5

The analogue signal produced by a transducer is sent by an ADC to a computer in the range 00H to FFH. If the set input is 70H calculate the fractional error when the output variable is (a) 80H and (b) 60H.

Solution
(a) Fractional digital error = (80H − 70H)/FFH = 10H/FFH = 0.10H.
(*Ans.*)
(b) Fractional digital error = (60H − 70H)/FFH = F0H/FFH = 0.F0H.
(*Ans.*)

The control algorithm is usually based upon difference equations with a sample time t because the analogue signal is sampled at regular t intervals (see Fig. 8.36). The proportional control signal $C = K_p E$ is replaced by $DC_n = K_p e_n$, where e_n is the nth sample of the error signal. The programming steps necessary are

```
Error = 0
Repeat
      time = 0
      read DCₙ from DAC
      read R from memory
      error = R − DCₙ
      DCₙ = Kₚ * error
      write DCₙ to DAC
      repeat
      until time = DAC sample time
Until switch-off
```

Digital proportional plus integral control

The algorithm for this control strategy is based upon the previous algorithm but with additional steps to give the integral control. The algorithm is

$$DC_n = K_p[e_n + (1/T_i)t(e_n + e_{n-1})/2]$$

The necessary programming steps for a computer to implement this algorithm are

```
Error = 0
Sum = 0
Repeat
    time = 0
    read DCn from DAC
    read R from memory
    error = R − DCn
    DCn = Kp * error
    Sum = Sum + (error + olderror)/2
    DCn = K[error + Sum * (sample time)/Ti)]
    write DCn to DAC
    olderror = error
    repeat
    until time = sample time
Until switch-off
```

Here Sum is the latest sum of the errors and T_i is the integral time.

Digital proportional plus integral plus derivative control

Lastly, the derivative action $C = T_d dE/dt$ in discrete form becomes

$$DC_n = T_d(e_n - e_{n-1})/t$$

Hence the algorithm for the third control strategy, i.e. PID, is

$$DC_n = K_p[e_n + (1/T_i)t(e_n + e_{n-1})/2 + (T_d/t)(e_n + e_{n-1})]$$

The programming steps for this algorithm are

```
Error = 0
Sum = 0
Repeat
    time = 0
    read DCn from DAC
    read R from memory
    error = R − DCn
    DCn = Kp * error
    Sum = Sum + (error + olderror)/2
    diff = error − olderror
    DCn = Kp[error + Sum * (sample time)/Ti)
        + diff * Td/(sample time)]
    write DCn to DAC
    olderror = error
    repeat
    until time = sample time
Until switch-off
```

Relative merits of analogue and digital control

(a) The real world is fundamentally analogue in nature, but the digital computer and the microprocessor employ digital techniques. This means that an ADC must be used to convert the analogue data into the digital format, and a DAC must be used to convert the digital data generated by the computer into analogue form. Each of these devices uses a number of bits to represent an analogue signal and so a loss of information must always occur. For example, if the temperature range from 0 to 100 °C is to be represented by 8 bits there can only be $2^8 = 256$ different temperatures represented. This means that temperature changes smaller than $100/256 = 0.39$ °C cannot be controlled. When the digital data is converted back into analogue form the signal can only contain 256 different voltage levels so that the output signal is not purely analogue.

(b) There is an inherent time difference between the real (analogue) world and the computer/microprocessor world. A practical control system may have to cater for time variations of from, perhaps, a few milliseconds to, more often, seconds and minutes, and sometimes hours or even days. On the other hand, a computer/microprocessor deals with events that last for just a few microseconds or nanoseconds. This means that if the value of an analogue quantity is supplied to a computer it must do so within a time period of about 1 μs. Sampling of the analogue input signals is therefore necessary.

(c) In the real world numerical information uses numbers to the base 10, i.e. 0, 1, 2, 3, etc. Computers and microprocessors, however, use numbers to the base 2 and so all numerical data require a number conversion.

(d) Digital control of a system is more flexible than analogue control because the control is vested in the easily reprogrammed software.

(e) Analogue systems generally have a better stability.

Instability in a control system

When the set input to a closed-loop control system and/or the load on the output of the system is suddenly changed the response of the system will depend upon its stability. If the system is stable the output quantity will change to its new value when the set input is changed,

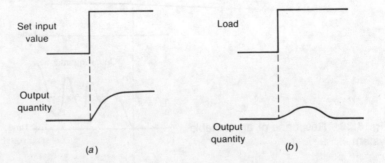

Fig. 8.37 Change in the output quantity of a stable system when a step is applied to (a) the input and (b) the load

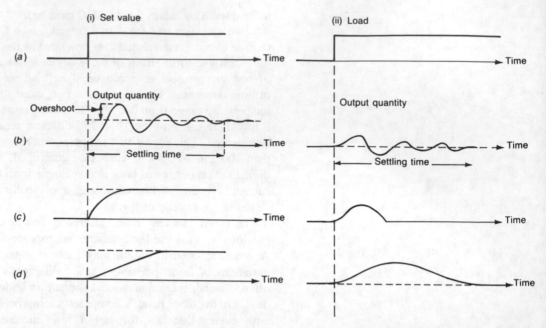

Fig. 8.38 (a) Step function applied to (i) input and (ii) load; (b) underdamped system; (c) critically damped system; (d) overdamped system

or will return to very nearly its old value when the load has changed, within a fairly short period of time. This desirable situation is illustrated by Fig. 8.37. The system is said to be stable because the output quantity remains under the control of the set input value. A system is *critically damped* if the change in the output quantity, or its return to its original value if the load is changed, takes place in the shortest possible time without any *overshoot*. Critical damping is shown by Fig. 8.38(c). Overshoot is the maximum difference between the transient and the steady-state responses of the system and it is illustrated by Fig. 8.38(b). This figure also illustrates the meaning of the term *underdamped*. When a system is underdamped the output quantity will both overshoot and oscillate about its desired steady-state value. The amplitude of the oscillation decreases with increase in time and the *settling time* of the system is the time that elapses before the output quantity reaches, and then remains within, a specified percentage of its final value. Usually this percentage is somewhere

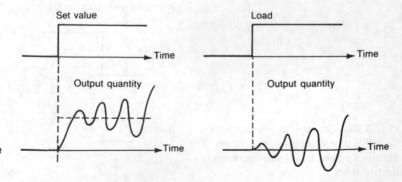

Fig. 8.39 Response of an unstable system

between 2 and 5. Lastly, the output quantity of an *overdamped* system, shown by Fig. 8.38(*d*), will neither overshoot its final value nor oscillate but it is slow to reach its steady-state value.

When a control system is unstable a change in either the set input value or the load will cause it to go into oscillation with an oscillation amplitude that increases with time until one of the devices in the system goes into saturation. Instability is illustrated by Fig. 8.39. A closed-loop system can only become unstable if its closed-loop gain becomes infinitely large. This situation will occur when the denominator in the closed-loop transfer function, equation 8.4, becomes equal to zero, i.e. when $GH = -1$. This means that a control system will be unstable if the open-loop transfer function GH has a gain of unity and a phase

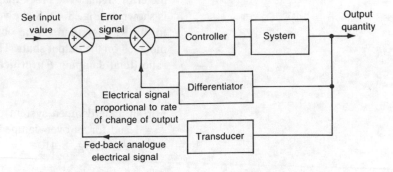

Fig. 8.40 Derivative damping

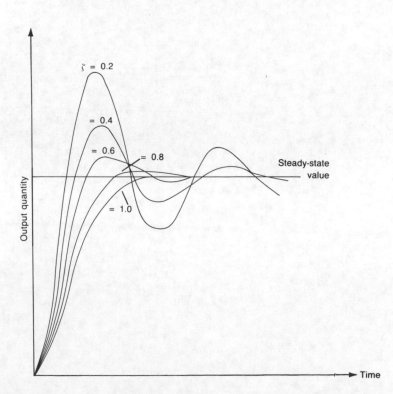

Fig. 8.41 Variation of the output quantity with time for various values of the damping ratio

shift of 180°. To avoid the possibility of instability in a control system it is necessary to design the system so that this condition can never exist.

A potentially unstable system is made stable by the addition of damping. Viscous damping is always present in any system because of the inevitable frictional losses. Extra viscous damping can be added to a position control system in the form of some kind of brake but not to a speed control system. Since viscous damping produces a power loss and the removal of the heat produced is a problem, it is only used in small RPC systems. The damping of a system can also be increased by the use of *derivative feedback*. This term means that a voltage that is proportional to the rate of change of the output quantity is fed back to the input of the system and here it is subtracted from the error signal. The block diagram of a *derivative damping* system is shown by Fig. 8.40. Since velocity $= d\theta_0/dt$ the damping voltage for an RPC system can be obtained from a tachogenerator that is mounted on the output shaft. The *damping ratio (coefficient)* ζ is the ratio (total damping F)/(critical damping F_c), i.e.

$$\zeta = F/F_c \tag{8.9}$$

For an underdamped system $\zeta < 1$, for critically damped system $\zeta = 1$ and for an over-damped system $\zeta > 1$. Some typical curves are given in Fig. 8.41.

9 Measurements

A wide variety of instruments are available for the measurement of electrical quantities such as current, voltage and power. The simpler, and more commonly employed, of these instruments are ammeters, voltmeters, both non-electronic and electronic, and the cathode ray oscilloscope (CRO). The basic principles of operation of these instruments have been discussed in the preceding volume and a knowledge of their operation is assumed in this chapter. The component values of capacitors and inductors can be measured by an a.c. bridge and their Q factor can be measured by a Q meter. A number of practical difficulties soon become apparent if an attempt is made to measure the signals within a digital system using an analogue instrument, such as a CRO, and digital instruments, such as the logic probe and the logic analyser, are often employed.

A number of terms are employed to describe the performance of a measurement instrument and the meanings of the more commonly employed are as follows:

1. The *sensitivity S* of an instrument is the ratio

$$S = \text{(change in scale indication)/(change in measured value)} \tag{9.1}$$

If, for example, an 0.5 V d.c. voltage is applied to the Y_1 input terminal of a dual-beam CRO and causes the visible trace to move through a distance of 1 cm the sensitivity of the CRO is 1/0.5 = 2 cm/V or 0.5 V/cm. For a Wheatstone, or a.c., bridge the sensitivity is expressed as

$$S = \text{(change in detector signal)/(change in input)} \tag{9.2}$$

The sensitivity of a non-electronic voltmeter is a means of expressing its input resistance and it is quoted in Ω/V. The input resistance of the voltmeter is equal to the product of its sensitivity and the full scale deflection (FSD) of the selected scale.

2. The *resolution* of an instrument is the smallest change in the measured value that will produce a change in the indicated value that can be seen. Often the resolution is expressed in terms of the maximum allowable input value. If, for example, a voltmeter has a fractional resolution of 2×10^{-4} and is used on its 0−2 V range then its resolution is $2 \times 2 \times 10^{-4} = 0.4\,\text{mV}$. If the meter is used on its

0–20 V range its resolution is $20 \times 2 \times 10^{-4} = 4\,\text{mV}$ and so on.

3. The *accuracy* of an instrument is the maximum amount by which its indicated value may differ from the true value. The accuracy may be either plus or minus the true value. The accuracy of a voltmeter is normally quoted in either one of two ways: (*a*) the accuracy of an analogue meter is quoted as a percentage of the FSD, and (*b*) the accuracy of a digital meter is quoted as a percentage of the displayed value ± one digit.

Example 9.1

A d.c. voltage is measured by (*a*) an analogue meter whose accuracy on its 0–30 V scale is ± 1% of FSD, and (*b*) a digital meter whose accuracy is ± 0.25% of read-out ± 1 digit. If the indicated voltage is 10.25 V determine the possible values of the measured voltage. Assume the digital meter to have a four-digit display.

Solution
(*a*) 1% of 30 V is 0.3 V and therefore the true value is between 10.25 ± 0.3 V = 9.95 V and 10.55 V.   (*Ans.*)
(*b*) 0.25% of 10.25 V is 0.0256 V and therefore the true value is between 10.25 ± 0.0256 ± 1 digit
$\qquad$ = 10.2244 − 0.01 or 10.21 V to 10.276 + 0.01 = 10.29 V.   (*Ans.*)

4. The *error* of a measurement is the difference between the indicated value and the true value. Usually, the error is quoted as a percentage of the true value.

Example 9.2

When a voltage of 10 V is measured by a voltmeter the indicated value is 10.51 V. Calculate the percentage error in the measurement.

Solution
% error = $[(10.51 - 10)/10] \times 100 = + 5.1\%$ (*Ans.*)

Example 9.3

At balance a Wheatstone bridge has its variable resistor R_1 set to 326.5 Ω and its two ratio resistors R_2 and R_3 both set to 3000 Ω. The tolerance of the variable resistor is ± 0.2% and the tolerance of the ratio resistors is ± 0.1%. Calculate the error in the measured value of the unknown resistance.

Solution
0.2% of 326.5 Ω is 0.65 Ω and so R_1 is either 325.85 Ω or it is 327.15 Ω; 0.1% of 3000 Ω is 3 Ω and so R_2 and R_3 are either 2997 Ω or 3003 Ω. Therefore
$\qquad R_{x(\text{max})}$ = (327.15 × 3003)/2997 = 327.81 Ω, and
$\qquad R_{x(\text{min})}$ = (325.85 × 2997)/3003 = 325.2 Ω.

The true value of R_x is 326.5 Ω. Hence
$$\% \text{ error} = [(327.81 - 326.5)/326.5] \times 100 = +0.4\%, \text{ or}$$
$$\% \text{ error} = [(325.2 - 326.5)/326.5] \times 100 = -0.4\%.$$
Therefore the percentage error = ±0.4%.   (*Ans.*)

The loading effect of a voltmeter

Whenever a voltmeter is connected across a part of a circuit to measure the voltage at that point the resistance of the voltmeter will shunt the circuit. This will alter the effective resistance across which the voltage being measured is dropped and so will alter the voltage. This *loading effect* will produce an error in the measured voltage.

Example 9.4

A voltmeter is used on its 0–100 V scale to measure the voltage dropped across the 47 kΩ resistor shown in Fig. 9.1. Calculate the percentage error in the indicated voltage if the voltage is (*a*) a d.c. voltage and (*b*) an a.c. voltage. The specification of the voltmeter includes: accuracy on d.c. ranges = ±1% FSD; accuracy on a.c. ranges = ±2% FSD; sensitivity on 100 V d.c. range = 20 kΩ/V; sensitivity on 100 V a.c. range = 2 kΩ/V.

Solution

The true voltage across the 47 kΩ resistor = $(200 \times 47)/147 = 63.95$ V.
(*a*) Meter resistance = $20 \times 10^3 \times 100 = 2$ MΩ and + 1% of FSD = + 1 V. This resistance is effectively connected in parallel with the 47 kΩ resistor to give a total resistance of
$$(47 \times 10^3 \times 2 \times 10^6)/(2 \times 10^6 + 47 \times 10^3) = 45.921 \text{ kΩ}.$$
The indicated voltage = $(200 \times 45.921)/145.921 = 62.94 \pm 1$ V. Thus the percentage error in the measured voltage is between
$$[(63.94 - 63.95)/63.95] \times 100 = -0.16\% \text{ and}$$
$$[(61.94 - 63.95)/63.95] \times 100 = -3.143\%   (Ans.)$$
(*b*) Meter resistance = $2 \times 10^3 \times 100 = 200$ kΩ and ± 2% FSD = ± 2 V. The effective parallel resistance of the 47 kΩ resistor and the voltmeter in parallel is $(47 \times 10^3 \times 200 \times 10^3)/(247 \times 10^3) = 38.057$ kΩ.
The indicated voltage = $(200 \times 38.057)/138.057 = 55.13 \pm 2$ V.
Therefore the percentage error is between
$$[(57.13 - 63.95)/63.95] \times 100 = -10.67\%, \text{ and}$$
$$[(53.13 - 63.95)/63.95] \times 100 = -16.92\%.   (Ans.)$$

It is clear from the previous example that to obtain a low loading error the voltmeter must have an input resistance that is much greater than the resistance across which it is connected. This is one reason why electronic voltmeters are so frequently employed. Nowadays nearly all the electronic voltmeters offered for sale are digital types and these, typically, have an input resistance of 10 MΩ or more on all voltage ranges. Most digital instruments are multimeters that are able to measure current and ohms as well as voltage. Figure 9.2 shows the basic block diagram of a digital multimeter.

If a current range is not available a digital voltmeter can be used to measure a current. A resistor, whose resistance is both small and

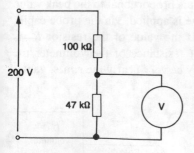

Fig. 9.1

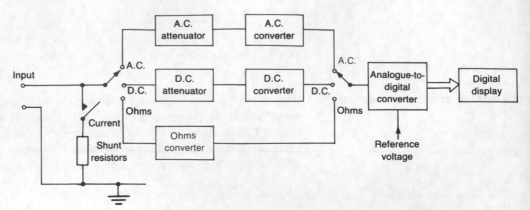

Fig. 9.2 Digital multimeter

accurately known, is connected in series with the circuit in which the current to be measured is flowing. The voltmeter is then used to measure the voltage across the resistor and then Ohm's law is applied to calculate the current. The resistor must be capable of carrying the current, be of stable value and have negligible self-capacitance and self-inductance.

Frequency effects on voltmeters

All meters, CROS and other measuring instruments have a particular range of frequencies over which they are able to operate satisfactorily. The moving-coil meter can only measure d.c. current and the moving-iron meter will only work up to about 100 Hz. Rectifier instruments that use a moving-coil meter can operate, according to the model, at frequencies up to about 20 kHz and in some cases even to about 100 kHz. The frequency range of an electronic meter, analogue or digital, is much wider and, again according to the model, may go as high as several megahertz. For measurements at radio frequencies specially designed RF voltmeters are available, some of which will operate up to several hundreds of megahertz and in a few cases to a few gigahertz.

The frequency range of any electronic voltmeter can always be extended with the aid of a *probe*. The circuit of a simple RF probe is shown by Fig. 9.3. It consists of an isolating capacitor, a low-capacitance diode and a resistor. The probe rectifies the RF input signal to produce a d.c. voltage that is proportional to the peak value of the input signal. The d.c. voltage is applied, via the probe cable, to the input of a d.c. voltmeter. If the value of the resistor R_1 is chosen appropriately for the input resistance of the voltmeter the voltage applied to the voltmeter may be very nearly the r.m.s. value.

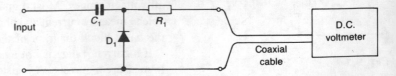

Fig. 9.3 RF probe

If, for example, the input resistance of the voltmeter is $10 \, M\Omega$ then choosing R_1 to be $4.7 \, M\Omega$ gives an input voltage of $V_m \times (10/14.7)$ $= 0.68 \, Vm$.

Waveform errors with voltmeters

A voltmeter is calibrated to indicate the r.m.s. value of a sinusoidal input voltage. Whenever the input voltage is of non-sinusoidal waveform there will be an error in the indicated value. Except for a relatively few true r.m.s. responding voltmeters, all voltmeters are of one of two basic types: (*a*) those in which the indication of the meter is proportional to the average value of a sinusoidal input voltage, and (*b*) those in which the indication of the meter is proportional to the peak value of a sinusoidal input voltage.

The basic arrangements of the two types of meter are shown by Figs 9.4(*a*) and (*b*). In a non-electronic voltmeter, like the AVOmeter, the a.c. amplifier in Fig. 9.4(*a*) is replaced by a transformer. Both types of meter are calibrated to indicate the r.m.s. value of an input sinusoidal waveform. This means that (*a*) an average-responding meter is calibrated to indicate 1.11 times the average value, and (*b*) a peak-responding meter is calibrated to indicate 0.707 times the peak value, of a sinusoidal input voltage. The calibrated indications of both types of meter remain accurate as long as the voltage measured is of purely sinusoidal waveform. Whenever, however, either type of meter is employed to measure a non-sinusoidal waveform there will be an error in the indicated value.

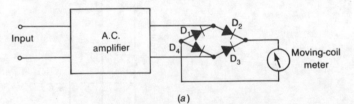

(*a*)

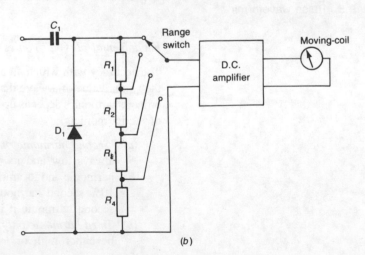

(*b*)

Fig. 9.4 A.C. voltmeters: (*a*) average responding; (*b*) peak responding

Average-responding meters

Square waveform

For a square waveform the average, r.m.s. and peak values are all equal to one another and hence an average-responding meter will indicate ·11% high. A square wave has the lowest ratio (r.m.s. value)/(average value) and so an average-responding meter will never indicate more than 11% high.

Pulse waveform

A pulse waveform is shown in Fig. 9.5. The width of each pulse is τ seconds and the time between the leading edges of the pulses — the periodic time — is T seconds. The duty factor (or cycle) of the waveform is the ratio τ/T and the average value of the waveform is $V_m(\tau/T)$. As the duty factor is reduced towards zero the voltage of each pulse must be increased by $1/\sqrt{(\text{duty factor})}$ to keep the r.m.s. value at a constant value. At the same time the average value is approaching zero. This means that an average-responding meter may indicate up to very nearly 100% low. However, for the majority of pulse waveforms the error is rarely in excess of 20% low.

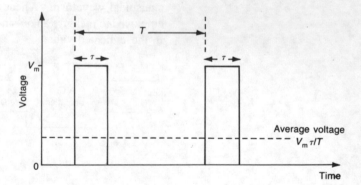

Fig. 9.5 Pulse waveforms

Fundamental plus harmonics

The accuracy with which an average-responding meter will indicate the r.m.s. value of a wave that contains both a fundamental and one or more harmonics depends upon the relative amplitude(s) and phase(s) of the harmonic(s):

(a) *Second harmonic only*: The error in the indicated value is always low and its value depends upon the order of the harmonic and its amplitude relative to the fundamental. For 10% second harmonic the error is 1% low and for 25% second harmonic it is 3% low.

(b) *Third harmonic only*: The error in the indicated value will be either high or low depending upon the phase of the harmonic, relative to the fundamental at time $t = 0$. For a third-harmonic content of 10% the error will $\pm 3.3\%$.

(c) *Several harmonics*: The indicated value is always lower than the true value.

Peak-responding meters

Square waveform

Since a square waveform has an r.m.s. value that is equal to its peak value the peak-responding meter will indicate 0.707 times the true r.m.s. value. The percentage error in the indicated value is hence

$$[(1 - 0.707)/1] \times 100 = 29.3\% \text{ low.}$$

Pulse waveform

Depending upon the duty factor the meter may indicate from 100% low to very high and no credence can be placed upon the measurement.

Fundamental plus harmonics

For the same harmonic content a peak-responding meter will always have a greater error than an average-responding meter.

Example 9.5

A rectifier instrument indicates a voltage of 20 V when it is used to measure (a) a square waveform and (b) a triangular waveform. For each wave calculate the peak and r.m.s. values and the percentage error in the indicated value.

Solution

The meter indicates 1.11 times the average value and hence the average value is $20/1.11 = 18.018$ V.

(a) For a square wave,

peak value = r.m.s. value = average value = 18.018 V.   (*Ans.*)

Percentage error = $[(20 - 18.018)/8.018] \times 100 = 11\%$ high.

(*Ans.*)

(b) For a triangular wave,

average value = (peak value)/2 and hence

the peak value = $2 \times 18.018 = 36.036$ V.

The r.m.s. value = $36.036/\sqrt{3} = 20.805$ V.   (*Ans.*)

Percentage error = $[(20 - 20.805)/20.805] \times 100 = 3.87\%$ low.

(*Ans.*)

The decibelmeter

A voltmeter can be calibrated to indicate decibel values directly and essentially a commercial decibelmeter merely consists of such a voltmeter. The voltmeter is used to measure the voltage dropped across a known value of resistance, generally $600\,\Omega$. This voltage will dissipate a power of $V^2/600\,\text{W}$ and this may be expressed in dBm (p. 123). Thus:

$$x\ \text{dBm} = 10\ \log_{10}[(V^2/600)/(1 \times 10^{-3})] \qquad (9.3)$$

The voltage that gives a power level of 0 dBm is easily found:

$$0\ \text{dBm} = 10\ \log_{10}[V^2/600)/(1 \times 10^{-3})]$$

Taking antilogs$_{10}$, or 10^x on a calculator,

$$1 = [V^2/600]/(1 \times 10^{-3})$$

so that
$$V = \sqrt{(600 \times 10^{-3})} = 0.775\ \text{V}.$$

Any or all of the scales on the voltmeter can be calibrated to indicate dBm. Suppose that the voltage scale chosen is the 0–10 V scale.

(a) 1 V: dBm $= 10\ \log_{10}[(1/600)/(1 \times 10^{-3})] = +2.22$ dBm.
(b) 2 V: dBm $= 10\ \log_{10}[(4/600)/(1 \times 10^{-3})] = +8.24$ dBm.

The remaining values calculated in the same way are shown in Table 9.1.

Other voltage scales can be similarly calibrated and Table 9.2 shows the figures that have been calculated for the voltage scale 0–1 V.

In a decibelmeter, or a voltmeter with a dB calibrated scale, the scale markings would be made for convenient decibel values as shown by Table 9.3.

A commercial decibelmeter will have all its scales marked in dBm values, but a voltmeter will probably have only one scale so marked. When a voltmeter is used to measure the decibel equivalent of a voltage outside of its calibrated scale a correction factor must be applied to the dBm reading. Consider the 0–1 V scale again and assume that

Table 9.1

Voltage (V)	1	2	3	4	5	6	7	8	9	10
dBm	2.22	8.24	11.76	14.26	16.2	17.78	19.12	20.28	21.3	22.22

Table 9.2

Voltage (V)	0.1	0.2	0.3	0.4	0.5	0.6	0.7	0.8	0.9
dBm	−17.78	−11.76	−8.24	−5.74	−3.80	−2.22	−0.88	+0.28	±1.30

Table 9.3

dBm	−6	−5	−3	−2	−1	0	+1	+2	+3	+4	+5
Voltage (V)	0.39	0.44	0.49	0.55	0.62	0.69	0.775	0.87	0.98	1.09	1.23	1.38

the scale markings for 0.1 V, 0.2 V, etc. correspond with the scale markings for 1 V, 2 V, etc. If the voltage is 0.5 V the pointer will lie over the 5 V mark on the 0–10 V scale and so it will indicate + 16.2 dBm. The correct dBm value is, from Table 9.2, −3.80 dBm. Thus, the indicated value is $16.2 - (-3.80) = 20$ dB high. Similarly, if the voltage is 0.2 V the indicated value is +8.24 dBm and the true value is −11.76 dBm. This is $8.24 - (-11.76) = 20$ dB high also. Hence the correction factor required is

$$CF = 20 \log_{10}[(\text{FSD of scale used})/(\text{FSD of calibrated scale})$$
$$(9.4)$$

When a voltmeter that has been calibrated to indicate decibel values is connected across a resistance of other than the calibration value an error in the indicated value will exist. If the calibration resistance is, as before, 600 Ω and the voltage is developed across a 1200 Ω resistance then the true dBm value is

$$10 \log_{10}[(1/1200)/(1 \times 10^{-3})] = -0.79 \text{ dBm}$$

but the indicated value is, from Table 9.1, +2.22 dBm. Hence the reading is $2.22 - (-0.79) = 3.01$ dB too high. This error is equal to

$$10 \log_{10}(1200/600) \text{ dBm}$$

In general, the resistance correction factor is

$$RCF = -10 \log_{10}(R/600) \text{ dB} \qquad (9.5)$$

where R is the value of the resistance across which the voltmeter is connected.

Example 9.6

A voltmeter has its 0–30 V scale calibrated to indicate dBm across 600 Ω. When the meter is used on its 0–3 V scale across a 100 Ω resistor the indicated dBm value is −2 dBm. Calculate the true dBm value.

Solution
From equations (9.4) and (9.5),
$$\text{true dBm} = -2 - 10 \log_{10}(100/600) + 20 \log_{10}(3/30)$$
$$= -2 + 7.78 - 10 = -4.22 \text{ dBm.} \quad (Ans.)$$

A.C. bridges

An *a.c. bridge* is employed to measure the value of either a capacitor or an inductor. The measurement is carried out by connecting the component under test, usually called the *unknown*, into one of the four arms of a bridge circuit. The other three arms of the bridge contain standard adjustable components whose values must be accurately known. One or two variable components are then adjusted until the bridge is *balanced*. Balance is indicated by zero current flowing in the detector. The use of a *null* method of measurement

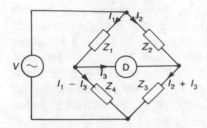

Fig. 9.6 Generalized a.c. bridge

has the advantage that its accuracy does not depend upon the calibration of an instrument but only upon the accuracy of the standard components that are employed and the sensitivity of the detector.

The generalized circuit of an a.c. bridge is given in Fig. 9.6 in which four impedances Z_1, Z_2, Z_3 and Z_4 are connected as shown. Applying Kirchhoff's voltage law to the circuit gives

$$V = I_1Z_1 + (I_1 - I_3)Z_4$$
$$= I_1(Z_1 + Z_4) - I_3Z_4 \tag{9.6}$$
$$V = I_2Z_2 + (I_2 + I_3)Z_3$$
$$= I_2(Z_2 + Z_3) + I_3Z_3 \tag{9.7}$$
$$0 = I_1Z_1 + I_3Z_D - I_2Z_2 \tag{9.8}$$

At balance $I_3 = 0$ and so equations (9.6) and (9.7) become

$$V = I_1(Z_1 + Z_4) \tag{9.9}$$

and

$$V = I_2(Z_2 + Z_3) \tag{9.10}$$

Therefore, $I_1(Z_1 + Z_4) = I_2(Z_2 + Z_3)$
or

$$I_1 = I_2[(Z_2 + Z_3)/(Z_1 + Z_4)] \tag{9.11}$$

Also equation (9.8) becomes $0 = I_1Z_1 - I_2Z_2$. Hence

$$I_1 = I_2Z_2/Z_1 \tag{9.12}$$

Equating equations (9.14) and (9.15) gives

$$(Z_2 + Z_3)/(Z_1 + Z_4) = Z_2/Z_1$$
$$Z_1Z_2 + Z_1Z_3 = Z_1Z_2 + Z_2Z_4$$

or

$$Z_1Z_3 = Z_2Z_4 \tag{9.13}$$

This means that the a.c. bridge is balanced when the products of the impedances of the opposite arms are equal to one another.

The detector used in an a.c. bridge is usually either a pair of earphones or an amplifier/detector whose output is applied to a centre-zero meter. To minimize errors caused by unwanted longitudinal earth currents between the voltage source and the bridge, a balanced and screened transformer is often employed to connect the voltage source to the bridge. It is advantageous if the leads that connect the unknown component to the bridge are as short as possible and, preferably, use screened cable. Since the value of a component and its self-resistance are to some extent frequency dependent, it is customary to measure a component at the frequency at which it is expected to be used. If the source frequency is then above the audio-frequency range the source voltage can be amplitude modulated and the detector can incorporate an AM detector stage.

A number of different a.c. bridge configurations have been developed and each has its own particular merits. It is an advantage if the balance conditions of the bridge do not include the frequency of the voltage source because it will then not be necessary for the source frequency to be accurately known, or for the supply to be a pure sinewave.

Inductance measurement

The value of an unknown inductance can be measured by balancing its impedance against the impedance of either a variable capacitor or a variable inductor. Variable inductors tend to be expensive and so a variable capacitor is most often employed.

The Maxwell bridge

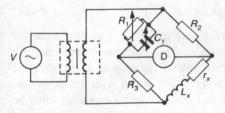

Fig. 9.7 Maxwell inductance bridge

The circuit of a Maxwell bridge is shown in Fig. 9.7. The bridge is suited to the measurement of the inductance and the self-resistance of a low Q factor inductor. It can be seen that the self-resistance is measured as a series-connected component. The impedance of the unknown inductance L_x and its self-resistance r_x is balanced against the total impedance of a variable capacitor C_1 in parallel with a variable resistor R_1. In balancing the bridge the values of C_1 and R_1 are alternately adjusted to reduce the current flowing in the detector D to the minimum possible value. The bridge has been balanced when the detector current has been reduced to zero. Then the product of the impedances of L_x and r_x in series, and of R_1 and C_1 in parallel, is equal to the product R_2R_3 of the two decade resistors. Then the value of the unknown component is (see Appendix A)

$$r_x = R_2R_3/R_1 \tag{9.14}$$
and
$$L_x = C_1R_2R_3 \tag{9.15}$$

Example 9.7

A Maxwell bridge is used to measure an unknown value of inductance. With the two decade resistors set to 1000 Ω the bridge is balanced when the variable capacitor is set to 2.5 µF and the variable resistor to 6422 Ω. Calculate the inductance and self-resistance of the inductance.

Solution
From equation (9.14), $r_x = (1 \times 10^6)/6422 = 155.7\,\Omega$. (*Ans.*)
From equation (9.15), $L_x = 1 \times 10^6 \times 2.5 \times 10^{-6} = 2.5\,H$. (*Ans.*)

The Hay bridge

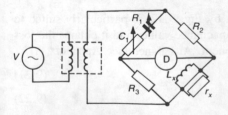

Fig. 9.8 Hay inductance bridge

The Hay bridge, shown in Fig. 9.8, is used to measure the value of a high-Q factor inductance and it gives the self-resistance of the inductance as an equivalent parallel resistor. At balance,

$$r_x = R_2R_3/R_1 \tag{9.16}$$
$$L_x = C_2R_1R_2 \tag{9.17}$$

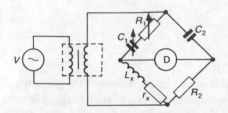

Fig. 9.9 Owen inductance bridge

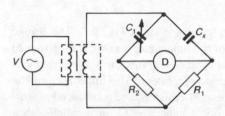

Fig. 9.10 Capacitance bridge

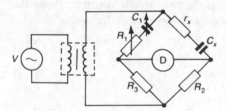

Fig. 9.11 De Sauty capacitance bridge

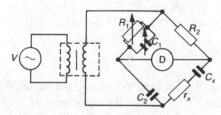

Fig. 9.12 Schering capacitance bridge

The Owen bridge

Another bridge that may be used to measure inductance is the Owen bridge shown in Fig. 9.9. The bridge is mainly used for the measurement of iron-cored inductors that may have to carry both d.c. and a.c. current.

Capacitance measurement

If the loss resistance of the capacitor under test is known to be negligibly small the simple bridge circuit given in Fig. 9.10 may be used. At balance the products of the impedances of the opposite arms are equal to one another. Hence

$$R_1/\omega C_1 = R_2/\omega C_x \quad \text{or} \quad C_x = C_1 R_2/R_1 \tag{9.18}$$

When the capacitor losses are not negligible they may be represented by an *equivalent loss resistance*. This loss resistance may be considered to be either a small-valued resistor in series with the capacitor, or a large-valued resistor in parallel with the capacitor. An a.c. bridge circuit that is to measure such a capacitor must be able to balance out both the capacitive reactance and the loss resistance and hence it must include a variable resistor as well as a variable capacitor.

The De Sauty bridge

The circuit of the De Sauty capacitance bridge is shown by Fig. 9.11 and it can be seen to obtain the loss resistance as a series component. At balance,

$$r_x = R_1 R_2/R_3 \tag{9.19}$$

$$C_x = C_1 R_3/R_2 \tag{9.20}$$

Because the resistance ratio needed to obtain resistance balance is the inverse of the resistance ratio needed for capacitance balance the value of the unknown capacitance must always be larger than the minimum value of the variable capacitor C_1. This difficulty is overcome by the Schering bridge.

The Schering bridge

The Schering bridge, shown by Fig. 9.12, is particularly suited to the measurement of small capacitance values and it obtains the loss resistance as a series component. At balance (see Appendix A)

$$r_x = C_1 R_2/C_2 \tag{9.21}$$

$$C_x = C_2 R_1/R_2 \tag{9.22}$$

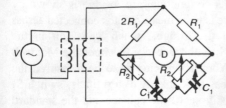

Fig. 9.13 Wien frequency bridge

Frequency measurement

If the balance conditions of an a.c. bridge include frequency then that bridge may be used to measure frequency. Although several bridge circuits could be used for this purpose the most commonly employed is the Wien bridge shown in Fig. 9.13. One of the decade resistors is exactly twice the resistance of the other. The required frequency range of the bridge is first selected by the setting of the ganged variable capacitor C_1 and then the ganged variable resistor R_2 is adjusted until the bridge is balanced. At balance the frequency of the voltage source is given by

$$f = 1/(2\pi C_1 R_2) \qquad (9.23)$$

To obtain an accurate result the voltage source should be filtered to remove any harmonic content that may be present.

Commercial a.c. bridges

A commercial a.c. bridge is usually of the *universal* type and so it is able to measure capacitance, inductance and resistance over a wide range of values within a specified range of frequencies. The bridge will have both an internal voltage source which may, or may not, be of variable frequency, and an input terminal that permits an external voltage source to be used when required. The bridge will also include its own detector and may sometimes provide the facility for an external detector to be used.

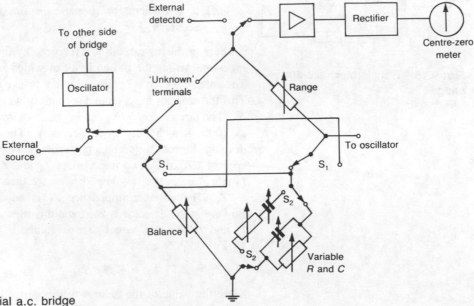

Fig. 9.14 Commercial a.c. bridge

The basic arrangement of a commercial a.c. bridge is shown by Fig.9. 14. When an inductance is to be measured it is connected across the 'unknown' terminals of the bridge, switch S_1 is left in the position shown, and switch S_2 is thrown to give either the Maxwell or the Hay bridge circuit. When a capacitor is to be measured the component is connected across the 'unknown' terminals, switch S_1 is operated to reverse the positions of the balance and the standard arms to give a capacitance bridge, and switch S_2 is put into the position that gives either a De Sauty or a Schering bridge as is required. To measure resistance the variable capacitors are switched out of circuit by switches that are not shown.

Such a bridge may be able to measure inductances between $0.2\ \mu H$ and $100\ H$, capacitances between $1\ pF$ and $1000\ \mu F$, and resistances between $10\ m\Omega$ and $10\ M\Omega$.

The transformer ratio-arm bridge

The conventional a.c. bridge circuits possess some disadvantages that are somewhat difficult and hence costly to overcome. These disadvantages are as follows:

(a) The balance conditions of a bridge rely upon the product or the ratio of two impedance standards. This limits its range of measurements because both low and high values of standard components are expensive to manufacture.

(b) The effects of unwanted stray capacitances and interference may often make it difficult to obtain a balance.

(c) For accurate measurements standards that cover several decades of resistance and reactance are necessary. Complex switching and wiring may then be required.

(d) It is difficult to measure *in situ* components with any accuracy.

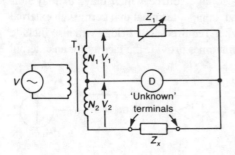

Fig. 9.15 Transformer ratio-arm bridge

These problems can all be overcome by the use of a *transformer ratio-arm bridge* the basic circuit of which is shown by Fig. 9.15. The voltage source is connected to the bridge via the transformer T_1. This transformer has a tapped secondary winding with turns N_1 and N_2. The turns ratio N_2/N_1 can be made as low as 1 : 1000 or as high as 1000 : 1 with very good accuracy. The voltages, V_1 and V_2, developed across the secondary windings N_1 and N_2 respectively, are applied across the two impedances Z_1 and Z_x. The currents I_1 and I_2 that flow in the two impedances are then $I_1 = V_1/Z_1$ and $I_2 = V_2/Z_x$. The variable impedance Z_1 is adjusted until the current flowing in the detector is zero and this means that the two currents I_1 and I_2 are then equal to one another. Therefore, at balance, $V_2/Z_x = V_1/Z_1$ and

$$Z_x = Z_1 V_2/V_1 = Z_1 N_2/N_1 \qquad (9.24)$$

Since the value of the unknown impedance is obtained in terms of both the standard impedance Z_1 and the turns ratio N_2/N_1 a wide

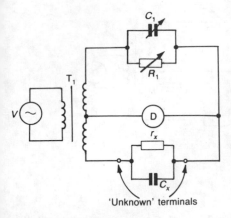

Fig. 9.16 Use of a transformer ratio-arm bridge to measure capacitance

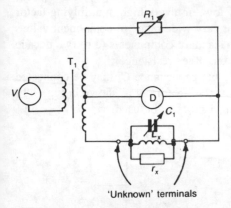

Fig. 9.17 Use of a transformer ratio-arm bridge to measure inductance

range of unknowns can be accurately measured with only a small range of standard impedances.

Measurement of capacitance

When capacitance is to be measured the capacitance is connected across the 'unknown' terminals and Z_1 consists of a variable capacitor C_1 in parallel with a variable resistor R_1. This is shown in Fig. 9.16 from which it can be seen that the equivalent parallel loss resistance of the capacitor is measured. At balance,

$$r_x = R_1 N_2 / N_1 \tag{9.25}$$

$$1/\omega C_x = (1/\omega C_1)(N_2/N_1) \quad \text{or} \quad C_x = C_1 N_1 / N_2 \tag{9.26}$$

When an inductance is to be measured the variable capacitance C_1 must be switched from the Z_1 position to be placed across the 'unknown' terminals. The arrangement is shown by Fig. 9.17. Capacitor C_1 is adjusted until it resonates with the unknown inductance L_x and this condition is indicated by the minimum current flowing through the detector. Once resonance has been obtained $\omega L_x = 1/\omega C_x$ and

$$L_x = 1/\omega^2 C_x \tag{9.27}$$

The variable resistor R_1 can then be adjusted to balance the bridge and then

$$r_x = N_2 R_1 / N_1 \tag{9.28}$$

Example 9.8

A transformer ratio-arm bridge with a transformer secondary ratio, N_1/N_2, of 4 is supplied at 1000 Hz and used to measure the value of (a) an unknown capacitor C_x, and (b) an unknown inductance L_x. When measuring C_x balance is obtained with the variable resistor R_x set to 98 kΩ and the variable capacitor set to 1380 pF. When measuring the inductance, balance is obtained with $R_1 = 10.42$ kΩ and $C_1 = 3970$ pF. Calculate the values of the unknown capacitance and inductance and their respective loss resistances.

Solution
(a) $C_x = 4 \times 1380 = 5520$ pF. (*Ans.*)
   $r_x = 98/4 = 24.5$ kΩ. (*Ans.*)
(b) $\omega L_x = 1/\omega C_x$,
   $L_x = 1/(4\pi^2 \times 1 \times 10^6 \times 3970 \times 10^{-12}) = 6.38$ H. (*Ans.*)

The range of a transformer ratio-arm bridge can be considerably increased if the detector is also supplied via a transformer as shown by Fig. 9.18. When the bridge is balanced there will be zero flux in the detector's transformer core and hence

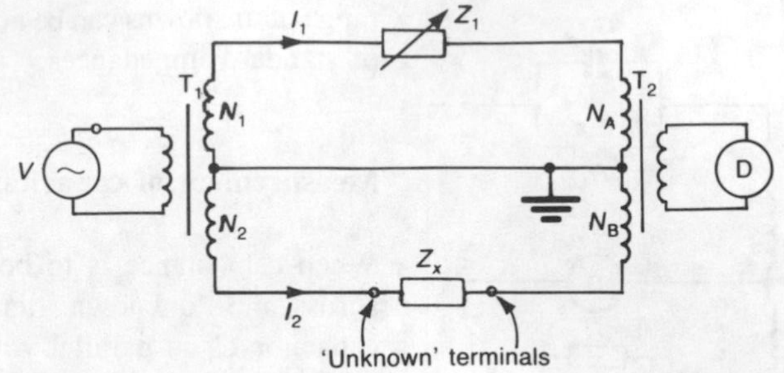

Fig. 9.18 Improved transformer
ratio arm bridge

$$I_1 N_A = I_2 N_B$$
$$V_1 N_A / Z_1 = V_2 N_B / Z_x$$

Therefore,

$$Z_x = V_2 N_B Z_1 / V_1 N_A = N_2 N_B Z_1 / N_1 N_A \qquad (9.29)$$

The unknown impedance Z_x is determined in terms of the standard impedance Z_1 and the product of two turns ratios N_2/N_1 and N_B/N_A. This means that either a very low, or a very high, multiplying factor can be obtained, permitting a very wide range of component values to be measured with just two standard components (e.g. two decade boxes, one capacitance and the other resistance).

The standard resistance R_1 and capacitance C_1 may be connected to separate tappings on the source transformer as shown in Fig. 9.19 to allow different multiplying factors to be used.

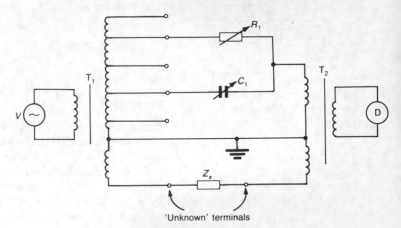

Fig. 9.19 How different multiplying
factors can be obtained

The advantages offered by the transformer ratio-arm bridge over the more conventional bridge are:

(a) Any unwanted impedances, due to stray or wiring capacitances for example, can be balanced out. This means that long leads can be used to connect the component under test to the bridge without the accuracy of the measurement being impaired.

(b) The multiplying factors provided by the transformer(s) are stable and do not vary with either time or temperature.

(c) Measurements can be made on components that are still connected in a circuit.

(d) The overall accuracy is good, being typically about 0.1%.

The Q meter

The Q factor of an inductor can be measured using an instrument known as a Q meter. The basic circuit of a Q meter is shown in Fig. 9.20. The inductor whose Q factor is to be measured is connected across the 'unknown' terminals of the meter and will then be in series with the variable capacitor C_1. The voltage source, an oscillator, provides a fixed-value constant current through the non-inductive resistor R_1. Here R_1 has a value in the region of 0.05 Ω and the voltage dropped across it is the voltage applied to the series-tuned circuit L_xC_1. This method of applying the voltage to the tuned circuit ensures that the effective source resistance of the supply is of very low value and so it does not reduce the Q factor of the circuit. The input voltage developed across R_1 is usually about 20 mV.

In the measurement of the Q factor of a component the frequency of the voltage source is set to the desired frequency of measurement and then capacitor C_1 is adjusted to give the maximum possible indication on the voltmeter. The L_xC_1 circuit is then series resonant and the voltage across C_1 is Q times greater than the voltage across R_1. Because the voltage across R_1 is kept at a constant value the voltmeter scale can be calibrated to indicate Q values. Strictly, the measured Q value is the Q factor Q_T of the series-tuned circuit. However, since $1/Q_T = 1/Q_L + 1/Q_C$, where Q_L is the Q factor of the inductor and Q_C is the Q factor of the capacitor, Q_T is very nearly equal to Q_L provided a high-quality, low-loss, variable capacitor is used.

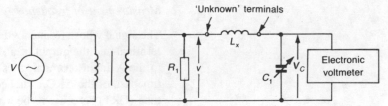

Fig. 9.20 *Q* meter

Example 9.9

An RF coil has an inductance of 50 μH and its Q factor at 3 MHz is to be measured by a Q meter. A maximum indication of 3.2 V is given by the voltmeter when the variable capacitor is set to 24.8 pF. If the effective source voltage is 0.05 V calculate (a) the Q factor of the inductor and (b) the self-resistance of the coil.

Solution
(a) $Q = [3.2/0.05] = 64$.   (*Ans.*)
(b) $r = \omega L/Q = (2\pi \times 3 \times 10^6 \times 50 \times 10^{-6})/64 = 14.73\,\Omega$.

(*Ans.*)

A Q meter can also be employed to measure capacitance. An inductor of suitable value is connected across the 'unknown' terminals of the meter and hence is in series with the variable capacitor. The variable capacitor is adjusted until it resonates with the inductor. Suppose that its resonant value is C_A. The capacitance to be measured is then connected in parallel with the variable capacitor. The circuit is then restored to its resonant condition by setting the variable capacitor to a new value C_B. The value of the unknown capacitor is then equal to $C_A - C_B$.

The cathode ray oscilloscope

The cathode ray oscilloscope (CRO) is an instrument which displays the variation with time of a signal that has been applied to one of its Y input terminals. The use of a CRO for testing and measurement has the advantage that it gives information about the signal waveform, but it is not as accurate as an electronic voltmeter for voltage measurements. Typically, the input impedance of a CRO is 1 MΩ in parallel with about 25 pF. The operation of a CRO and its use for voltage and period (and hence frequency) measurements has been discussed in the preceding volume.

Lissajous' figures

A *Lissajous' figure* is a pattern that is displayed by a CRO when sinusoidal voltages are simultaneously applied to both a Y input and the X input terminals with the internal timebase switched OFF. The pattern can be used to measure both frequency and phase.

Measurement of frequency

The signal whose frequency is to be measured is applied to a Y input terminal and the output of a variable-frequency oscillator is applied to the X input terminal, as shown by Fig. 9.21. With the internal timebase of the CRO switched OFF the display on the cathode ray tube (CRT) screen will be a pattern that depends upon the ratio of the two frequencies. A stable pattern is obtained for certain frequency ratios such as 2 : 1, 3 : 1 and 3 : 2, and these patterns are shown by Fig. 9.22. The ratio of the number of times the pattern is tangential

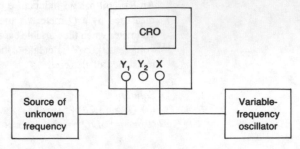

Fig. 9.21 Measurement of frequency using Lissajous' figures

(a)　　　　　(b)　　　　　(c)

Fig. 9.22 Lissajous' figures for frequency ratios of (a) 2 : 1; (b) 3 : 1; (c) 3 : 2

to the horizontal to the number of times it is tangential to the vertical is equal to the ratio of the two frequencies.

Example 9.10

The Lissajous' figure method is employed to measure the frequency of a signal. When the variable-frequency oscillator has been set to 1500 Hz the displayed pattern is shown by Fig. 9.22(*b*). Calculate the frequency of the signal.

Solution
$f = 1500 \times 3/2 = 2250$ Hz.　　(*Ans.*)

Measurement of phase

When the two signals applied to the Y and X inputs of a CRO are sinusoidal and at the same frequency the display will be one of the patterns given in Fig. 9.23. The circular pattern will only appear if the two voltages applied to the X and the Y deflecting plates are of equal amplitude. The phase angle θ between the two signals can be determined from the pattern. Referring to Fig. 9.24, the angle θ is given by equation (9.30), i.e.

$$\theta = \sin^{-1}(a/b) \tag{9.30}$$

$\theta = 0°$ or 360°　　$\theta = 45°$ or 315°　　$\theta = 90°$ or 270°　　$\theta = 135°$ or 225°　　$\theta = 180°$

Fig. 9.23 Lissajous' pattern for frequency ratio of 1 : 1

This equation can be applied whatever the relative amplitudes of the two signals may be. The method is of limited accuracy because of the difficulties that are experienced in (*a*) ensuring that the centre of the ellipse is accurately positioned on the graticule, and (*b*) measuring *a* and *b* accurately.

Phase measurement

The phase difference between two sinusoidal signals at the same frequency can be measured using a CRO. One method that employs

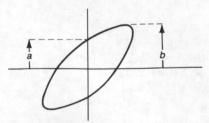

Fig. 9.24 Phase measurement using Lissajous' figures

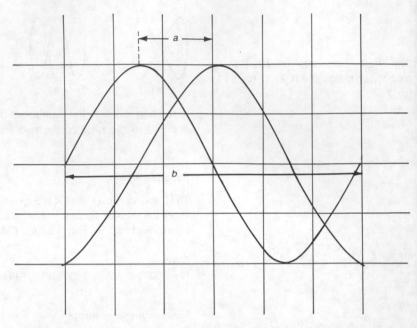

Fig. 9.25 Phase measurement

Lissajous' figures has already been discussed and two other methods are available.

1. With the internal timebase switched ON the two signals are applied to the Y_1 and Y_2 input terminals of the CRO. The displayed waveforms will then be similar to those shown in Fig. 9.25. Measurements must then be made, using the graticule on the face of the screen, of the periodic time b of the waveforms and of the time a between the positive peaks of the two waveforms. Then the phase difference θ between the two signals is

$$\theta = 360a/b \qquad (9.31)$$

Example 9.11

In the measurement of the phase difference between two sinusoidal signals the display shown in Fig. 9.25 was obtained. If the volts/centimetre switch was set to 3 V/cm and the timebase switch was set to 200 μs/cm calculate (*a*) the r.m.s. voltage, (*b*) the frequency of each wave and (*c*) the phase difference between them.

Solution
(*a*) Peak-to-peak voltage = 4 cm = 4 × 3 = 12 V.
Therefore r.m.s. voltage = 6 × 0.707 = 4.242 V.   (*Ans.*)
(*b*) Periodic time T = 6 cm = 6 × 200 = 1200 μs.
Therefore frequency = $1/T$ = 833.3 Hz.   (*Ans.*)
(*c*) Phase difference = θ = 360° × 1.5/6 = 90°.   (*Ans.*)

2. With the internal timebase switched on, one signal is connected to the Y_1 input terminal and the other signal is connected to the Y_2

terminal. The channel selector is then switched to A + B to display the sum waveform and its peak-to-peak value V_S is noted. The channel selector is then switched to A − B to display the difference waveform and its peak-to-peak value V_D is noted. The phase difference θ between the two signals is then given by

$$\theta = 2\,\tan^{-1}(V_D/V_S) \tag{9.32}$$

Measurement of an amplitude-modulated wave

If the envelope of an AM wave is to be displayed on a CRO the AM wave is applied to a Y input terminal and the timebase is set to the frequency of the modulating signal. The modulation envelope will then be visible but the modulated carrier itself will appear as just a blur. If there is a need to examine the carrier itself the timebase frequency must be set to be equal to the carrier frequency, but it will then not be possible to observe the modulation envelope.

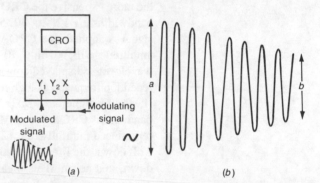

Fig. 9.26 Measurement of an AM waveform: (*a*) connection; (*b*) display

An alternative method, which makes it much easier to determine the modulation factor of the AM wave and/or to detect any distortion that may be present, is shown by Fig. 9.26(*a*). The AM wave is applied to a Y input terminal and the modulating signal is applied to the X input terminal. The internal timebase is switched OFF. The display on the CRT screen is both stationary and of trapezoidal shape, as shown by Fig. 9.26(*b*). To determine the modulation factor m of the AM wave the distances a and b are measured and then

$$m = (a - b)/(a + b) \tag{9.33}$$

Measurement of rise time

When a square, or a rectangular, waveform is displayed on the screen of a CRO the timebase should be adjusted to give a stationary display. The peak voltage and the periodic time can then be easily measured. The measurement of the rise time is a little more difficult. The leading

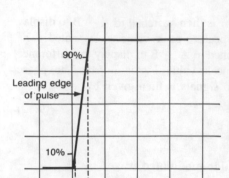

Fig. 9.27 Measurement of rise time

Instruments for digital measurements

edge of the pulse waveform should be displayed by adjustment of both the timebase and the trigger controls. The volts/centimetre control can then be adjusted to give a reasonable height to the display and then the Y shift control is used to make the top and bottom of the display line up with two horizontal graticule lines. This makes it easier to estimate the 10 and 90% points on the leading edge of the pulse. The X shift control is then used to position the 10% point on to a vertical graticule line and, finally, the distance to the 90% point is measured. The idea is illustrated in Fig. 9.27. The X distance can then be converted into time by multiplying it by the time/centimetre setting of the timebase control.

Bandwidth limitations of CROs

The bandwidth of a CRO refers to the frequency at which the gain of each Y amplifier has fallen by 3 dB from its low-frequency value. The bandwidth varies considerably with the type of CRO; in general the more expensive the CRO the wider will be its bandwidth. Typical bandwidths are 12, 20, 40, 50 and 100 MHz. If a 12 MHz sinusoidal signal is applied to a CRO with a 12 MHz bandwidth the displayed amplitude will be only 70.7% of the true amplitude. If a pulse waveform is displayed considerable distortion may occur if its pulse repetition frequency is high relative to the 3 dB bandwidth of the CRO. If, for example, a 4 MHz square wave is applied to a 12 MHz bandwidth CRO, the 4 MHz fundamental component will be reproduced faithfully, the 12 MHz third-harmonic component will be 3 dB down, the fifth-harmonic component will be approximately 6 dB down, and so on. This means that the displayed signal will not be of square waveform. Even worse, if the frequency of the square wave were equal to the bandwidth of the CRO the attenuation suffered by the harmonics would mean that a somewhat distorted sinusoidal waveform would be displayed.

The CRO is an instrument that works in the time domain and it is triggered by a single event in the input signal. The timebase is synchronized by trigger pulse and this can often mean, when displaying a digital pattern, that the display is continually changing and it is often impossible to decipher. Most CROs have, at most, just two channels and this number is not sufficient for the testing of many digital systems which employ buses with a width of 8, 16 or even 32 bits. Even if a digital pattern can be observed it is difficult, if not impossible, for the observer to identify a single error in a stream of data. Furthermore, the signals in a digital system mean very little if seen in isolation and it is really necessary for all the signals to be monitored together. The CRO is therefore unsuited to the testing of, and to fault location on, high-speed digital systems and either the *logic analyser* or the *signature analyser* is employed.

Analysers are expensive pieces of equipment and not the simplest to operate, and for many tests simpler test devices such as the *logic pulser*, and the *logic probe* are employed.

The logic pulser

The logic pulser is a device that is used to apply a logical signal to a pin on a digital IC package. When it is non-operative the logic pulser has a high output impedance. To apply a signal to an IC the pulser is held firmly against the appropriate pin and then a button is pressed. This action causes the pulser to emit one or more narrow pulses, (typically $0.8-1.8\,\mu$s), going first high and then low, into the IC pin. A switch on the pulser allows the emitted pulse waveform to be a single pulse, four pulses or a continuous stream of pulses. The source, or sink, current is large enough to ensure that the emitted pulse(s) control the operation of the circuit. No damage is caused to the circuit under test because the pulse(s) is/are of such narrow width. The output of the circuit can then be checked with the aid of a logic probe.

The logic probe

A logic probe is a circuit, powered by the IC under test, that when applied to an output of the IC under test is able to detect, memorize and display the logic level(s) at that point. The basic logic probe, shown in Fig. 9.28, has two LED indicators, one green and the other red. If the IC pin tested is at the logical 1 level the circuit sinks current and the red LED glows visibly. If the IC pin is at the logical 0 level the circuit sources current and the green LED glows. More complex logic probes incorporate extra circuitry to provide more features:

(a) A pulse stretcher will allow single short pulses to be detected.
(b) The output of the probe may be buffered to prevent the probe loading the circuit under test.
(c) In transistor–transistor logic (TTL) circuits a floating input will give a logical 1 indication but some logic probes are able to indicate a floating input.
(d) An indication may be given of the nature of a pulse waveform at the tested point.

The indications given by one type of logic probe are listed in Table 9.4.

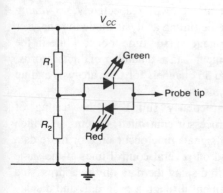

Fig. 9.28 Logic probe

The logic checker

The *logic checker* is a device that can be clipped on to an IC and then it will indicate the logical state of each pin. The device uses red LEDs as the indicators and these glow visibly to indicate logic 1 and do

Table 9.4

Red LED	Green LED	Yellow LED	State
ON	OFF	OFF	Logic 1
OFF	ON	OFF	Logic 0
OFF	OFF	OFF	Open circuit test point
ON	ON	Flashing	Square wave, $f < 1$ MHz
OFF	OFF	Flashing	Square wave, $f > 1$ MHz
ON	OFF	Flashing	Logic 1 pulses
OFF	ON	Flashing	Logic 0 pulses

not glow to indicate logic 0. The logic checker is only suitable for use on a static, or a low-frequency, digital system since only then will the observer be able to follow the changing indications.

The logic analyser

A logic analyser is a test instrument that is used to capture and trace groups of digital signals obtained from between 8 and 48 inputs. The inputs are sampled in sequence and the samples are stored in a *data acquisition memory*. A logic analyser can be used in either one of two modes of operation that are known, respectively, as *timing analysis* and *state analysis*. In the timing analysis mode the analyser displays the input data in a voltage versus time mode, while in the state analysis mode it displays input data as a number of digital words. Typically, there might be up to 36 channels for time analysis and up to 72 channels for state analysis.

The logic analyser is particularly suited to the testing of microprocessors and microprocessor-controlled systems in which changes in the logical state of signals may occur on 8 or more data lines, on 16 address lines, and on several control lines at the same time. The instrument is able to display the data on the address and the data lines, trace the movement through a program, and display all the instructions that have been implemented.

The logic analyser can be used to analyse the operation of both the hardware and the software of the system tested. Most logic analysers are flexible in that their functions are programmable and they can be used to analyse the signals originated by any microprocessor. A logic analyser is connected to the circuit under test by means of an interface module that is commonly known as a 'pod'. The main functions of a pod are threefold:

(a) to synchronize the analyser to the system under test;
(b) to format the system data into the form that the analyser is able to understand;
(c) to provide all the necessary electrical and mechanical connections.

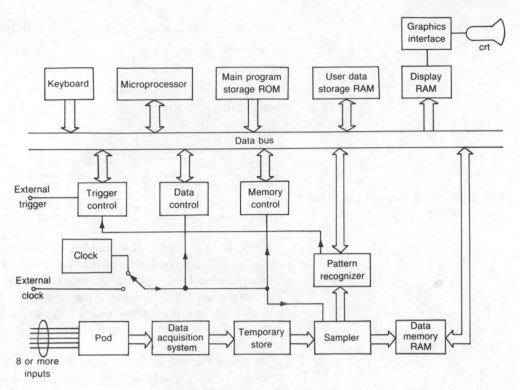

Fig. 9.29 The logic analyser

The basic block diagram of a logic analyser is shown in Fig. 9.29. The data originating from eight or more sources is applied via the pod to the data acquisition circuit. The acquisition of data is initiated by a trigger signal and the data is held in a temporary store. The trigger may be provided externally or it may be initiated by the pattern recognizer recognizing a selected word in the captured data. The contents of this store are sampled at regular intervals and the samples are stored in the data RAM. Once a complete set of sampled data has been stored the microprocessor will command the data control, and the memory control, circuits to transfer the data to the display RAM and thence to the graphics interface. The graphics interface IC controls the display that is visible on the screen of the CRT.

The operation of the logic analyser is controlled by means of a Qwerty keyboard. A menu appears on the screen to give guidance to the operator on various set-up choices he can make and this often includes a 'Help' option.

Timing analysis

In the timing analysis mode of operation of a logic analyser the input signals are displayed on the screen as functions of time. The mode employs an internal clock that is not synchronized with the clock in the system under test. This allows the data to be sampled at a faster rate than would otherwise be possible and so allows all data, including

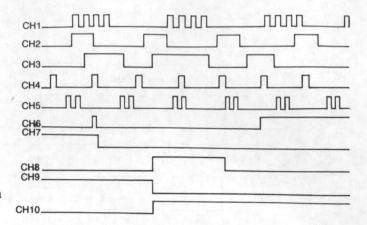

Fig. 9.30 Time analysis display of a logic analyser

glitches, to be recorded. The samples of each signal voltage are compared with a threshold value held in memory and each sample is interpreted as being at either the logical 1 or the logical 0 level. The data displayed on the screen is therefore a 'clean' digital waveform with no noise or distortion and with all transitions from 0 to 1, and from 1 to 0, clearly shown. A typical time analysis display is shown in Fig. 9.30.

To ensure that no data is lost and that the level transitions are precisely located the sampling rate should be set to at least ten times the reciprocal of the expected minimum pulse width. Thus, if the minimum pulse width is 100 ns the sampling frequency should be at least 100 MHz.

Timing analysis is used to detect hardware errors in the system under test that will cause errors in timing. It allows the operator to see where transitions occur in a data stream, whether or not particular signals

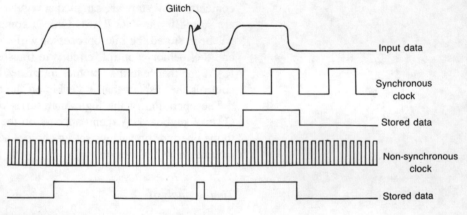

Fig. 9.31 Glitches in the data stream are not detected by synchronous sampling but are detected by non-synchronous sampling

arrive at the correct instants in time, and whether or not any voltage spikes exist (see Fig. 9.31).

State analysis

A logic analyser can also display its input data in the form of a list or a graph. The displayed information may include the data held at various addresses in the system tested, the program flow using assembly language, and statistical information such as how often specific code segments are used or the average execution time for different instructions. The state analyser samples signals from the system under test using the system's own clock and it only looks at the states of the inputs at the trailing edge of each clock pulse. Any narrow pulses or glitches that may occur in between successive clock pulses are not detected (see Fig. 9.31).

The list and graphs that are generally available are as follows:

State list. The state list displays data in binary, hexadecimal, octal, decimal or ASCII (American Standard Code for Information Interchange) form. The information is presented in a number of groups of individual signals under labels that show the values of groups of lines for various sample numbers. A typical screen display is shown by Fig. 9.32.

STATE	BIN	ST	ADR	DATA
−0001	1101001	0110011	000550	CFCB
TRIG	1101001	0110011	000552	0000
0001	1101001	0110011	000554	F866
0002	1101001	0110011	000556	F436
0003	1101001	0110011	000558	61A2
0004	1101001	0110011	000560	0000
0005	1101001	0110011	000562	5641
0006	1101001	0110011	000564	564B
0007	1101001	0110011	000566	564C
0008	1101001	0110011	000568	564D
0009	1101001	0110011	000570	580C

Fig. 9.32 State analysis: state list

	ADDR	INST	Mnemonic
0000	3000	9680	LDA A 80
0001	3002	900A	SUB A 0A
0002	3004	9782	STA A 82
0003	3006	9690	LDA A 90
0004	3008	48	ASL A
0005	3009	48	ASL A
0006	300A	9B90	ADD A 90
0007	300C	9B82	ADD A 82
0008	300E	9782	STA A 82
0009	3010	3F	SW1

Fig. 9.33 State analysis: disassembly list

Disassembly list. The state analyser incorporates a disassembler to make the programme flow in the microprocessor under test easier to follow. A typical screen display is shown by Fig. 9.33.

State graph display. When the logic analyser is operated in this mode the contents of the analyser's memory are displayed in graphical form.

Some logic analysers use a split screen to allow both timing and state information to be displayed at the same time. In the testing of a microprocessor system the operator of a logic analyser will first employ state analysis to check the overall software performance. If the program is not running correctly the state analysis will allow the faulty area to be localized and then timing analysis will be used to locate the faulty IC.

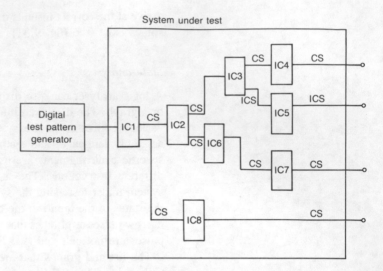

Fig. 9.34 Use of a signature analyser: CS = correct signature; ICS) incorrect signature

The signature analyser

A signature analyser is an instrument that is used in fault-finding which compresses long streams of data into short test results that are known as signatures. A digital test pattern is supplied to the input of the system under test and a probe is used to sample the signals that then exist at various points in the system. The results are displayed by a CRT and if the digital test pattern — the signature — is correct the probe is moved to another point further along the signal path. This procedure is followed repeatedly until a faulty signature is found and this is the point where the fault lies. The idea is illustrated in Fig. 9.34 in which IC 3 is supposed to be faulty. Obviously, it is necessary to know what the correct signature is at each point in the system and this information is usually given in the handbook for the equipment.

Signal flow in a measurement system

A measurement system is an arrangement of circuits and devices that performs a measurement function. The signal to be measured is applied to the input of the measurement system and there is then a signal flow through the system to the point where an analogue indication or a digital read-out is provided. The signals are of various magnitudes, waveforms and types and in their passage through the measurement system may be changed one or more times. A transducer may change the signal from one form to another, an amplifier will alter the amplitude of the signal, an ADC will change an analogue signal into digital form, a rectifier will change an a.c. signal into a d.c. signal and so on. All these processes are not perfect; they may introduce distortion, have inadequate bandwidth, be non-linear, etc. and will, in some way, degrade the signal a little and, in addition, the signal may be subject to noise and interference.

At the output of a measurement system the results have often to be read and interpreted by a human operator. Here such factors as fatigue, mental and physical, boredom, speed of response, skill, etc. come into play and affect the overall accuracy of a measurement.

Appendix A

The balance equations of an a.c. bridge can be derived using the operator j and two examples are given.

(a) The Maxwell bridge: at balance,

$$[r_x + j\omega L_x][R_1/(1 + j\omega C_1 R_1)] = R_2 R_3$$

$$r_x R_1 + j\omega L_x R_1 = R_2 R_3 + j\omega C_1 R_1 R_2 R_3$$

Equating real terms:

$$r_x R_1 = R_2 R_3 \quad \text{or} \quad r_x = R_2 R_3/R_1. \tag{9.14}$$

Equating j terms:

$$L_x R_1 = C_1 R_1 R_2 R_3 \quad \text{or} \quad L_x = C_1 R_2 R_3. \tag{9.15}$$

(b) The Schering bridge:

$$[R_1/(1 + j\omega C_1 R_1)][r_x + 1/(j\omega C_x)] = R_2/j\omega C_2$$

$$r_x R_1 + R_1/j\omega C_x = R_2[1 + j\omega C_1 R_1)/j\omega C_2$$

$$j\omega C_2 R_1 r_x + R_1 C_2/C_x = R_2 + j\omega C_1 R_1 R_2$$

Equating real parts:

$$R_1 C_2/C_x = R_2 \quad \text{or} \quad C_x = R_1 C_2/R_2 \tag{9.22}$$

Equating j parts:

$$C_2 R_1 r_x = C_1 R_1 R_2 \quad \text{or} \quad r_x = C_1 R_2/C_2 \tag{9.21}$$

Exercises

Chapter 1

1.1 A 1000 Ω resistor is connected in parallel with a 2.2 μF capacitor across a 20 V, 400 Hz supply. Calculate (a) the current flowing into the circuit and (b) the impedance and admittance of the circuit.

1.2 A coil of inductance 0.25 H and resistance 200 Ω is connected in parallel with a 220 nF capacitor across a 25 V, 400 Hz supply. Calculate (a) the impedance and (b) the power factor of the circuit.

1.3 An inductance of 0.2 H and 100 Ω resistance is connected in series with a capacitor of 1 μF. The current that flows in the circuit is 15 mA at 1 kHz. Calculate (a) the voltage across, and the impedance of, the circuit, (b) its power factor and (c) the power dissipated.

1.4 A 5 mH inductor L_1 is connected in series with a 300 Ω resistor. This combination is then connected in parallel with another 5 mH inductor L_2. The self-resistances of the two inductors are negligibly small. At a certain frequency f the voltage across the circuit is 10 V and the current flowing in the resistor is 20 mA. Calculate (a) the frequency of the supply and (b) the total current taken from the supply.

1.5 A circuit consists of an inductor of 0.05 H and resistance 5 Ω in parallel with a 0.1 μF capacitor. Calculate (a) the resonant frequency, (b) the Q factor and (c) the dynamic resistance of the circuit.

1.6 An a.c. voltage of 10 V is connected across a circuit that consists of a 100 Ω resistor in series with a 0.22 μF capacitor. Determine the frequency at which the phase angle between the supply voltage and the current is 45°.

1.7 A 2.2 μF capacitor is connected in parallel with a 100 Ω resistor and the combination is connected in series with a 47 Ω resistor. When a voltage source at frequency $5000/2\pi$ Hz is connected across the circuit the current in the 100 Ω resistor is 10 mA. Calculate (a) the total current, (b) the applied voltage, (c) the power dissipated in the circuit and (d) the power factor of the circuit.

1.8 Two loads are connected in parallel to a 240 V, 50 Hz single-phase supply: (i) 3 kW at 0.6 lagging power factor; (ii) 2 kW at unity power factor. Calculate (a) the overall power factor of the system, (b) the capacitance that must be connected in parallel with the loads to give an overall power factor of unity and (c) the total current taken from the supply before and after the capacitance is connected.

1.9 A parallel-resonant circuit consists of a coil of 15 mH inductance and 10 Ω resistance in parallel with a capacitor of 220 nF. Calculate (a) the resonant frequency, (b) the dynamic resistance, (c) the 3 dB bandwidth of the circuit and (d) the supply current if the capacitor current is 20 mA at the resonant frequency.

1.10 An electrical motor takes a current of 12 A at a lagging power factor of 0.5 from a 240 V, 50 Hz supply. Calculate (a) the parallel capacitance needed to improve the power factor to 0.91 and (b) the total current taken from the supply after power factor correction.

1.11 A 100 mH inductor has a self-resistance of 30 Ω and it is connected in parallel with a 22 μF capacitor. A 200 V, 50 Hz voltage supply is connected across the circuit. Calculate (a) the total current, (b) the circuit's admittance and impedance and (c) the power dissipated.

Chapter 2

2.1 Three 48 Ω resistors are star-connected to a three-phase supply. If the line voltage is 415 V calculate (a) the line current and (b) the total power dissipated.

2.2 Three equal loads, each having $R = 20$ Ω and $X_L = 20$ Ω, are star-connected across a 415 V, 50 Hz three-phase supply. Calculate (a) the line currents and (b) the total power dissipated.

2.3 Three 2.2 kΩ resistors are delta-connected to a 415 V, 50 Hz three-phase supply. Calculate (a) the line and phase currents and (b) the total power dissipated.

2.4 A line voltage of 415 V is applied to a balanced delta load. The total power dissipated is 15 kW at a power factor of 0.7 lagging. Calculate the line and phase currents.

2.5 Three 300 Ω resistors are connected (i) in star, and (ii) in delta to a three-phase 415 V, 50 Hz supply. Calculate (a) the total power dissipated and (b) the line current for each connection. Comment on the results.

2.6 Three identical inductors are connected in star and take a total power of 4.5 kW at a power factor of 0.2 lagging from a 415 V three-phase supply. Calculate (a) the line currents and (b) the resistance and inductance of each inductor.

2.7 A balanced load takes a total power of 10 kW at a power factor of 0.6 lagging from a power supply. The power is measured using the two-wattmeter method. Calculate the indications of the two wattmeters.

2.8 An unbalanced star-connected load that consists of three resistors of 100, 200 and 300 Ω respectively is supplied by a four-wire 415 V, 50 Hz three-phase supply. Calculate the current that flows in the neutral conductor.

2.9 The phase voltages in a four-wire three-phase system are 240 V. The three load impedances are (*a*) a 200 Ω resistance in series with an inductive reactance of 1000 Ω, (*b*) a 400 Ω resistor and (*c*) a 330 Ω resistor. Calculate the total power dissipated in the load and the overall power factor.

Chapter 3

3.1 A voltage source has an internal resistance of 10 Ω and an e.m.f. of 1 V. It is connected via a transformer to a resistive load of 100 Ω. Determine the turns ratio of the transformer that will give the maximum load power and the value of this power.

3.2 Determine the Thevenin equivalent circuit for the network shown in Fig. E.1. Then calculate the current flowing in a 100 Ω resistor connected across the output terminals of the network.

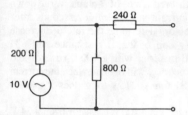

Fig. E.1

3.3 Apply Thevenin's theorem to the circuit given in Fig. E.2 to find the voltage across the capacitor.

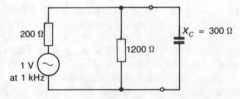

Fig. E.2

3.4 For the circuit given in Fig. E.3 calculate (*a*) the value of load resistance that will dissipate the maximum power and (*b*) the value of this power.

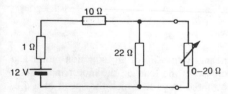

Fig. E.3

3.5 Use (*a*) Thevenin's theorem, (*b*) Norton's theorem and (*c*) the superposition theorem to calculate the load current I_L in the circuit shown in Fig. E.4.

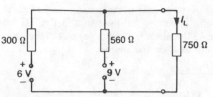

Fig. E.4

3.6 A transformer is to be used to match a 50 Ω, 3 V a.c. voltage source to a 200 Ω resistive load. Calculate (*a*) the required transformer turns ratio, (*b*) the load power and (*c*) the turns ratio that would give the maximum load power if the load had an inductive reactance of 50 Ω in series with the 200 Ω resistance.

3.7 A voltage source is connected to a purely resistive load. The variation of the load current I_L with the load resistance R_L is shown by Table E.1. Assuming that the internal impedance of the source is purely resistive, determine the Thevenin and Norton equivalent circuits of the source.

Table E.1

$R_L (\Omega)$	200	400	700	1200
$I_L (mA)$	12	10	8	6

3.8 The voltage source shown in Fig. E.5 has an e.m.f. of 4 V and, at a frequency of 159 Hz, an internal impedance that consists of 600 Ω resistance in series with a capacitance of 1.25 μF. Calculate the components of the load impedance Z_L that will give the maximum load power. Also, calculate the value of this maximum load power.

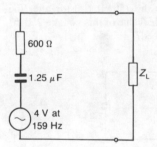

Fig. E.5

3.9 For the network shown in Fig. E.6 use (*a*) Thevenin's theorem and (*b*) Norton's theorem to find the maximum load power.

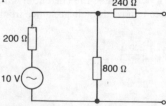

Fig. E.6

3.10 Calculate the turns ratio of the transformer in Fig. E.7 for the maximum power to be dissipated in the 2 Ω load resistance if the 42 Ω resistor is taken as (a) part of the source or (b) part of the load on the source.

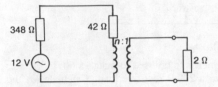

Fig. E.7

Chapter 4

4.1 In the circuit shown in Fig. E.8 the capacitor C is initially discharged. At the instant the switch S is closed the current I is 500 μA. The current then falls exponentially to 400 μA. Find the time constant of the circuit. Calculate also the values of the two resistors R_1 and R_2.

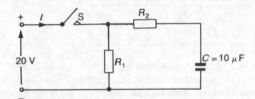

Fig. E.8

4.2 If the capacitor in Fig. E.9 is initially discharged calculate the voltage across the 150 kΩ resistor after the switch S is closed. What is the value of this voltage 250 ms after the switch is then opened?

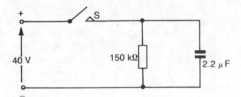

Fig. E.9

4.3 For the circuit given by Fig. E.10 determine (a) the time constant, (b) the initial current and (c) the final current flowing in the circuit. (d) What will be the current 2.7 ms after the switch is operated to remove the voltage source from the inductance?

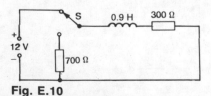

Fig. E.10

4.4 For the circuit given in Fig. E.11 calculate (a) the time constant on charge, (b) the initial current, (c) the final current, (d) the capacitor voltage when fully charged and (e) the energy stored once steady-state conditions prevail.

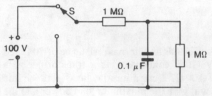

Fig. E.11

4.5 Derive an expression for the fall time of a rectangular pulse in terms of the time constant of an $R-C$ circuit.

4.6 A constant voltage is maintained across an inductance of L henrys in series with a resistance of R ohms. Write down the expression for the current in the circuit t seconds after switching ON. What factors determine the rate of increase of the current? A relay coil of 200 Ω resistance and 8 H inductance is connected in series with a 100 Ω resistor and a 60 V voltage source. The relay will operate when the current flowing in its coils is 31.6 mA. How long does it take to operate?

4.7 A capacitor is charged via a 100 kΩ resistor from a 24 V d.c. voltage source and the capacitor voltage has risen to 12 V in 10 ms. Calculate the capacitance of the capacitor. Sketch the current/time and capacitor voltage/time curves for this circuit.

4.8 An inductor has an inductance of 20 H and a resistance of 100 Ω. The inductance is switched across a 50 V d.c. voltage source. Calculate (a) the initial current, (b) the initial rate of change of the current, (c) the rate of change of the current when the current is 0.2 A and (d) the final current.

4.9 A d.c. voltage of 30 V is applied to an $R-C$ circuit. The steady-state condition is reached after 7.5 s and the energy stored in the capacitor is 9.9 mJ. Calculate the values of the capacitance and the resistance of the circuit.

4.10 A rectangular waveform of periodic time 10 ms is applied to an $R-C$ circuit and the output voltage is taken from the resistor. Estimate what the time constant of the circuit should be if (a) the waveform is to be unchanged, (b) the waveform is to be differentiated and (c) the waveform is to be integrated.

Chapter 5

5.1 What is meant by the slip of an induction motor and why does it increase as the load on the motor increases?

5.2 Explain the three ways in which the speed of a d.c. motor can be controlled. Why is it usually better to control the speed by adjusting the field current than by adjusting either the armature current or the supply voltage?

5.3 A three-phase motor produces an output power of 2 kW when it is supplied with a balanced input at 0.82 power factor. If the line voltage is 415 V and the line current is 4 A calculate

(a) the input power and (b) the efficiency of the motor.

5.4 What factors determine the output torque of a d.c. motor? A d.c. motor has an armature that is 15 cm long and has a diameter of 10 cm. The armature winding has 100 turns and the magnetic field strength is 50 mT. Calculate the average torque produced when the average armature current is 4.5 A.

5.5 A d.c. shunt-connected generator has a field resistance of 150 Ω and an armature resistance of 2 Ω and supplies 1 kW power to a 25 Ω load resistance. The total losses in the generator can be represented by a 120 Ω resistor connected across the output terminals. Calculate the overall efficiency of the generator.

5.6 A 300 V shunt-connected d.c. generator has an armature resistance of 0.25 Ω and a field resistance of 100 Ω. The armature rotates at 600 r.p.m. to supply a 2 kW load. At what speed must the armature rotate for the motor to supply a 3 kW load? Assume that the field current remains at a constant value.

5.7 Why is a starter needed for a d.c. motor? A 250 V d.c. series motor takes a current of 25 A from the voltage supply to produce an output torque of 25 N m at a speed of 750 r.p.m. Calculate (a) the current taken from the supply and (b) the speed of rotation when the output torque is increased to 50 N m. The total resistance of the armature and the field winding is 0.8 Ω.

5.8 A machine produces a mechanical output power of 20 kW with 82% efficiency. If the iron losses are 1100 W and the copper losses are 1900 W, calculate the friction/windage loss.

5.9 A d.c. shunt motor has an armature resistance of 0.1 Ω and runs at 1000 r.p.m. off a 300 V supply when the armature current is 100 A. Calculate its speed when the load torque is doubled.

Chapter 6

6.1 Express each of the following powers in dBm: (a) 12 mW, (b) 33 mW and (c) 1.5 W. What will be the new value of each if the power is (i) doubled, and (ii) halved?

6.2 An amplifier has an input resistance of 2000 Ω and a load resistance of 1200 Ω. When the voltage across the input terminals of the amplifier is 100 mV the load voltage is 2.5 V. Calculate the gain of the amplifier in dB.

6.3 The band of frequencies 100 Hz to 20 kHz is applied to the input of the cascaded filter network shown in Fig. E.12. Determine the frequencies that appear at the output of the system.

Fig. E.12

6.4 A band-pass filter characteristic is often obtained by connecting a low-pass filter and a high-pass filter in cascade.

Determine the cut-off frequencies of the two filters required to produce a pass bandwidth of 4–7 kHz. Discuss (a) why the loss in the passband is not zero for an *LC* filter, and (b) the advantages of active filters over *LC* filters.

6.5 Design a T-attenuator to have a characteristic resistance of 600 Ω and an attenuation of 9 dB. The attenuator is placed at the input of an amplifier that has a gain of 17 dB. Calculate the output voltage of the amplifier in (a) volts and (b) dBm when the input voltage to the attenuator is 0.4 V. The input and load resistances of the amplifier are both equal to 600 Ω.

6.6 Define the decibel. The input voltage to a 10 dB attenuator is at a level of −20 dB relative to 1 mV. Calculate the output voltage of the attenuator in (a) dB relative to 1 μV and (b) volts.

6.7 The input signal to a 20 dB attenuator varies between 23.5 mW and 1.25 W. (a) Express each power in dBm and state the fluctuation in the signal level in dB. (b) State the maximum and the minimum output powers in dBm and their difference in dB.

6.8 An $R-C$ filter has a 3 dB cut-off frequency of 3000 Hz. Calculate the frequency at which the output voltage is 6 dB down.

6.9 An attenuator has a characteristic resistance of 600 Ω and it is connected between matched source and load resistances. The input voltage is +16 dB relative to 1 mV and the attenuator has a loss of 12 dB. Calculate (a) the input and output voltages and (b) the output power.

6.10 (a) Calculate the power values of (i) 20 dbW, (ii) −10 dBW and (iii) +3 dBm. (b) An amplifier has a gain of 26 (times) when its output terminals are connected to the input of a 10 dB attenuator. Calculate the necessary input voltage to the amplifier for the attenuator output voltage to be 50 mV. Assume that all inputs and outputs are of the same resistance.

Chapter 7

7.1 Explain what is meant by FSK. Calculate the bandwidth required for an FSK system if the bit rate is 2400 b/s. The bandwidth of a commercial-quality speech circuit is 300–3400 Hz so why cannot the PSTN transmit this FSK signal? What other form of digital modulation is employed?

7.2 Draw the basic block diagram of a PCM system and give the signal waveform at the output of each block. What is meant by the term 'quantization' and why is it used? What quantum levels do each of the following binary numbers represent? (a) 11111111, (b) 00100000, (c) 10000111 and (d) 00001010.

7.3 The instantaneous voltage of an AM wave is given by $v = 120(1 + 0.6 \sin 3000\pi t) \sin (2\pi \times 10_r^6)$ V. Determine (a) the peak unmodulated carrier voltage, (b) the carrier frequency, (c) the modulating frequency, (d) the modulation factor and (e) the modulating signal voltage.

7.4 What is meant by (a) modulation of a carrier wave and (b) demodulation of a carrier wave? A sinusoidal 60 V carrier is modulated by a ± 10 V square wave (a) in amplitude, (b) in frequency and (c) in phase. Draw the modulated wave in each case.

7.5 A sinusoidally modulated AM wave has a maximum value of 75 V and a minimum value of 25 V. Draw the waveform. Calculate (a) the unmodulated carrier voltage, (b) the modulating signal voltage, (c) the upper side frequency voltage and (d) the modulation factor.

7.6 A carrier wave is frequency modulated using a modulator whose sensitivity is 500 Hz/V. The modulating signal is a 2.5 kHz sine of amplitude 4 V. (a) Calculate the bandwidth occupied by the modulated wave. (b) What would this bandwidth become if the modulating signal had its frequency doubled and its voltage halved?

7.7 The instantaneous current of an AM wave is given by
$$i = 5 \sin (2\pi \times 10^4 t) +$$
$$1.5 \cos (2\pi \times 0.95 \times 10^4 t) -$$
$$1.5 \cos (2\pi \times 1.05 \times 10^4 t) \text{ amperes}$$
(a) Determine the carrier frequency and the frequency of the modulating signal. (b) Calculate the total power dissipated when the current flows in a 200 Ω resistor. (c) Calculate the percentage (side frequency power)/(total power) and comment on the result.

7.8 Discuss the advantages of FM over AM. Why is FM not used for medium-wave sound broadcasts? The RF bandwidth required by a FM system is 100 kHz when the modulation index is 4 rad. If the modulating signal voltage is increased by 6 dB, calculate (a) the new modulation index and (b) the new bandwidth.

7.9 An FM system transmits the commercial-quality speech bandwidth of 300–3400 Hz using a carrier frequency of 100 MHz and a deviation ratio of 15. A 12 V, 3400 Hz audio-frequency signal produces the rated system deviation. Calculate (a) the rated system deviation, (b) the frequency deviation produced by a 10 V, 2000 Hz modulating signal, (c) the modulation index produced by a 6 V, 1000 Hz modulating signal and (d) the minimum bandwidth that must be provided for the system.

7.10 A 2000 Hz sine wave of amplitude 8 V is transmitted over a PCM system. The sampling frequency is 8000 Hz. Calculate (a) the instantaneous voltage of the baseband signal at each sampling instant, assuming that the first sample is taken when time $t = 0$ and (b) the binary train transmitted to line if the system employs 128 quantum levels.

Chapter 8

8.1 The devices in the forward path of a closed-loop control system have the following transfer functions: amplifier, 20 V/V; voltage/frequency converter, 50 Hz/V; electric motor, 500 r.p.m./Hz; electric pump, 4 litres/1000 r.p.m.; water tank, 1 metre/2000 litres. Determine the overall transfer function of the forward path.

8.2 A d.c. generator voltage reference circuit uses an amplifier with a transfer function K_A, a d.c. generator that generates 250 V per 1 A field current and a d.c. reference voltage of 85 V. The steady-state terminal voltage of the generator is to be only 0.1 V less than the no-load value when a current of 20 A is taken by the load. Calculate the necessary values for the amplifier gain K_A and the fraction K_F of the

terminal voltage that is fed back and compared with the reference voltage.

8.3 Draw the block diagrams of (a) an open-loop system and (b) a closed-loop system and list their relative merits. State with explanations which of the systems listed below are open-loop systems: (a) a refrigerator, (b) a television receiver, (c) an ammeter and (d) an electric iron.

8.4 Determine the transfer function of the circuit given in Fig. 8.19 when the input signal is a step function.

8.5 What decides whether a control system is of the first or of the second order? Draw graphs to show the expected transient and steady-state responses of a first-order control system. Show the effect of altering the time constant of the system.

8.6 What is meant by (a) a tachogenerator and (b) a servomotor? Draw graphs to illustrate the meanings of the terms overdamped, critically damped, and underdamped when applied to a closed-loop control system. Which of these systems is likely to become unstable? How is the risk of instability reduced? The total viscous damping in a control system is 50 N m per rad/s. If the damping ratio is 0.7 calculate the value of the critical damping.

8.7 Draw the block diagram of a proportional control system. What improvements in the behaviour of the system are obtained by the addition of (a) integral control and (b) derivative control? Amend the block diagram to show how these controls are applied. How can the stability of the system be improved without an increase in viscous damping? Draw a third block diagram to show how this other kind of damping is applied to the system.

8.8 State the three main control strategies and show how they can be implemented using (a) analogue and (b) digital techniques. Briefly discuss some of the problems that may occur when a process is controlled by a computer. Why is it that the computer deals with samples derived from the output quantity?

8.9 With the aid of a block diagram explain how a digital computer can be used to control the temperature and pressure of a liquid within a large tank as well as the opening and closing of input and output valves to control the flow of the liquid. The set temperature is 7AH and the input range is from 00H to FFH. Calculate the digital error and the fractional error for a digital signal representing the actual temperature of C4H.

8.10 A system controls the filling of bottles with a liquid. The sequence of events is as follows: check if a bottle is in the correct position; if yes then stop the conveyor belt; start to fill the bottle, when full turn off the flow of the liquid; move filled bottle underneath capping machine and cap; feed finished bottle to output; repeat the sequence until a stop signal is given. Draw a suitable flowchart for this operation and draw a block diagram of the system.

Chapter 9

9.1 A transformer ratio-arm bridge has a turns ratio $N_2 = 2N_1$ and is supplied by a 10 kHz voltage source. Balance is obtained when the variable resistor $R_1 = 1.5\,\text{k}\Omega$ and the variable capacitor $C_1 = 0.18\,\mu\text{F}$. Calculate the values of (a) a capacitor, and (b) an inductor connected across the 'unknown' terminals.

9.2   List the advantages of a logic analyser over a CRO for the testing of a digital system. Briefly explain why a synchronous clock is used for state analysis and a non-synchronous clock for timing analysis. The timing analyser of a logic analyser is to be able to capture 20 ns glitches. Calculate the minimum clock frequency required.

9.3   A CRT has a Y plate sensitivity of 60 V/cm and an X plate sensitivity of 65 V/cm. A 50 V, 10 kHz sinusoidal signal is applied to the Y plates and a 260 V peak-to-peak, variable frequency sawtooth voltage is applied to the X plates. The frequency of the sawtooth voltage is varied until a stationary display of five cycles is seen. Calculate the height and the width of the displayed waveform.

9.4   (a) Describe how a simple electronic voltmeter can be used to measure in turn voltage, current and power. (b) What are the likely sources of error in these measurements? (c) How are the results likely to be affected by the waveform of the signal being measured?

9.5   Explain why errors will occur when a non-sinusoidal signal is measured by an electronic voltmeter. An electronic voltmeter indicates a voltage of 6.66 V. What is the percentage error in the indicated value when the voltmeter is (a) average responding, and (b) peak responding, and the measured voltage is of square waveform?

9.6   An AM wave is to be measured by connecting it to the Y input of a CRO. State, with explanations, whether the timebase should be synchronized to the carrier frequency or to the modulating signal frequency. Draw the displayed waveform if its depth of modulation is 30%. Also draw the waveform that would be displayed if the internal timebase of the CRO were switched OFF and the modulating signal were applied to the X input terminal.

9.7   Describe the operation of a Q meter for the measurement of the Q factor of (a) an inductor and (b) a capacitor. In a measurement of the Q factor of a capacitor a standard inductance of $50\,\mu\text{H}$ was used. Without the capacitor connected to the Q meter resonance was obtained with $C_1 = 200\,\text{pF}$ and $Q = 85$. With the capacitor connected resonance was obtained with $C_1 = 100\,\text{pF}$ and $Q = 82$. Calculate the capacitance and the Q factor of the capacitor.

9.8   Draw the circuit of any a.c. bridge that can be used to measure the inductance and self-resistance of an inductor. State whether the self-resistance is obtained as a series- or a parallel-connected component. What are the advantages offered by a null method of measurement? Why may it sometimes be difficult to balance the bridge? How can this difficulty be partly, at least, overcome?

9.9   The voltage between the points B and C in Fig. E.13 is to be measured using a non-electronic voltmeter having a sensitivity of 3 kΩ/V. Determine the percentage error in the measurement if the voltmeter is connected (a) between the points B and C and (b) between the points A and B. The meter is used on its 0–10 V range.

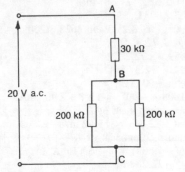

Fig. E.13

9.10   (a) A digital voltmeter has a quoted accuracy of ±0.1% + 1 digit. Calculate the limits of the voltage when the displayed value is 20.42 V. (b) A decibelmeter has been calibrated using a 600 Ω resistance. Calculate the true value if the meter indicates +2.5 dBm when it is connected across a 1200 Ω resistor. (c) What determines whether an a.c. bridge can also be used to measure frequency?

Answers to numerical exercises

1.1 (a) $I_R = V/R = 20/1000 = 20$ mA.
$X_C = 1/(2\pi \times 400 \times 2.2 \times 10^{-6})$
$= 180.9\,\Omega,$
$I_C = 20/180.9 = 110.6$ mA.
$I = \sqrt{(20^2 + 110.6^2)} \angle \tan^{-1}(-110.6/20)$
$= 112.4 \angle -79.8°$ mA.
(b) $Z = V/I = 20/(112.4 \times 10^{-3}) \angle -79.8°$
$= 178 \angle 79.8°\,\Omega.$
$Y = 1/Z = 5.62 \angle -79.8°$ mS.

1.2 (a) $X_L = 628.3 \angle 90°\,\Omega.$
$Z_L = \sqrt{(200^2 + 628.2^2)} \angle \tan^{-1}628.3/200$
$= 659.4 \angle 72.3°\,\Omega.$
$I_L = 25/659.4 \angle 72.3°$
$= 37.9 \angle -72.3°$ mA.
$X_C = 1808 \angle -90°\,\Omega.$
$I_C = 25/1808 \angle -90° = 13.83 \angle 90°$ mA.
The in-phase component of I_L
$= 37.9 \cos(-72.3°) = 11.52$ mA.
The quadrature component of I_L
$= 37.9 \sin(-72.3°) = -36.11$ mA.
$I_L - I_C = -22.28$ mA.
Hence I
$= \sqrt{(11.52^2 + 22.28^2)} \angle \tan^{-1}(-22.28/11.52)$
$= 25.08 \angle -62.7°$ mA.
Impedance $Z = 25/(25.08 \times 10^{-3} \angle -62.7°)$
$= 996.8 \angle 62.7°\,\Omega.$
(b) Power factor $= \cos(-62.7°) = 0.46.$

1.3 (a) $V_R = IR = 0.015 \times 100 = 1.5$ V.
$V_L = IX_L = 0.015 \times 2\pi \times 1000 \times 0.2$
$= 18.85$ V.
$V_C = IX_C = 0.015/(2\pi \times 1000 \times 1 \times 10^{-6}) = 2.39$ V.
$V = \sqrt{[1.5^2 + (18.85 - 2.39)^2]} = 16.53$ V.
$Z = V/I = 16.53/0.015 = 1102\,\Omega;$
$\phi = \tan^{-1}(16.46/1.5) = 84.8°.$
(b) Power factor $= \cos 84.8° = 0.09.$
(c) Power dissipated $= VI \cos 84.5°$
$= 22.32$ mW.

1.4 (a) $V_R = 300 \times 0.02 = 6$ V.
$V_{L_1} = 2\pi f \times 5 \times 10^{-3} \times 0.02$
$= 6.283 \times 10^{-4}f$ V.
$10^2 = 6^2 + V_{L_1}^2$, so $V_{L_1} = 8$ V.
Therefore, $f = 8/(6.283 \times 10^{-4})$
$= 12.733$ kHz.
(b) $I_{L_2} = 10/(2\pi \times 12.733 \times 10^3 \times 5 \times 10^{-3})$
$= 25$ mA.

$\phi_{L_1} = \tan^{-1}(V_{L_1}/V_R) = \tan^{-1}(8/6) = 53.1°.$
In-phase component of $I_{L_1} = 20 \cos 53.1°$
$= 12$ mA.
Quadrature component of $I_{L_1} = 20 \sin 53.1°$
$= 16$ mA.
$I = \sqrt{[12^2 + (25 + 16)^2]}$
$= 42.72 \angle 73.7°$ mA.

1.5 (a) $1/LC = 2 \times 10^8$ and $R^2/L^2 = 1 \times 10^4$ and hence
$f_0 = 1/(2\pi\sqrt{LC}) = 2251$ Hz.
(b) $Q = \omega_0 L/r = (2\pi \times 2251 \times 0.05)/5$
$= 141.4.$
(c) $R_d = Q\omega_0 L = 141.4 \times 2 \times \pi \times 2251 \times 0.05 = 100$ kΩ.
Or $R_d = L/Cr = 0.05/(0.1 \times 10^{-6} \times 5)$
$= 100$ kΩ.

1.6 $\tan 45° = 1 = X_C/100 = 1/200\pi fC,$
and $f = 1/(200\pi \times 0.22 \times 10^{-6}) = 7234$ Hz.

1.7 $V_C = 100 \times 0.01 = 1$ V.
$X_C = 1/(5000 \times 2.2 \times 10^{-6}) = 90.9\,\Omega.$
$I_C = V/X_C = 1/90.9 = 11/90°$ mA.
(a) $I = \sqrt{(10^2 + 11^2)} \angle \tan^{-1}(11/10)$
$= 14.87 \angle 47.7°$ mA.
(angle relative to V_R)
(b) Voltage across $47\,\Omega = 14.87 \angle 47.7° \times 47 = 0.7 \angle 47.7°$ V.
Take I as the reference phasor, and resolve 1 V into its horizontal and vertical components:
$1 \cos 47.7° = 0.673$ V and
$1 \sin 47.7° = 0.74$ V. Therefore
$V = \sqrt{[(0.7 + 0.673)^2 + 0.74^2]} = 1.56$ V.
$\phi = \tan^{-1}(0.74/1.373) = 28.3°.$
(c) $P_{100} = 0.01^2 \times 100 = 10$ mW.
$P_{47} = (14.87 \times 10^{-3})^2 \times 47 = 10.4$ mW.
Therefore the total power dissipated
$= 20.4$ mW.
(d) Power factor $= P/VI = 20.4/(1.56 \times 14.87)$
$= 0.88.$

1.8 (a) Total real power $P = 3 + 2 = 5$ kW.
$\phi = \cos^{-1}0.6 = 53.1°.$
Total reactive volt-amps $Q = 3 \tan 53.1°$
$= 4$ kvar.
Therefore the apparent power $S = \sqrt{(P^2 + Q^2)}$
$= \sqrt{(5^2 + 4^2)} = 6.4$ kVA.
Overall phase angle $= \tan^{-1}(Q/P) = \tan^{-1}(4/5)$
$= 38.7°.$
Hence the overall power factor $= \cos 38.7°$
$= 0.78$ lagging.

(b) For unity power factor the parallel capacitor must pass a current I_C of 4000/240
= 16.67 A.
Therefore, $X_C = V/I_C = 240/16.67 = 14.4 \, \Omega$
$= 1/100\pi C$,
or $C = 221 \, \mu F$.

(c) Original supply current = 6400/240
= 26.67 A.
Current after power factor correction
= 5000/240 = 20.83 A.

1.9 (a) $f_0 = 1/[2\pi\sqrt{(15 \times 10^{-3} \times 220 \times 10^{-9})}]$
= 2.77 kHz.

(b) $R_d = L/Cr$
$= (15 \times 10^{-3})/(220 \times 10^{-9} \times 10) = 6818 \, \Omega$.

(c) $Q = \omega_0 L/r = (2\pi \times 2.77 \times 10^3 \times 15 \times 10^{-3})/10 = 26.1$.
$B_{3dB} = f_0/Q = 2770/26.1 = 106.1$ Hz.

(d) $I_C = QI$, $I = 20/26.1 = 0.767$ mA.

1.10 (a) The real power is to remain the same. Therefore,
$P = 12 \text{ V} \times 0.5 = VI_C \times 0.91$. Hence,
$I_C = (0.5 \times 12)/0.91 = 6.593$ A.
$\phi_L = \cos^{-1}0.5 = 60°$.
$\phi = \cos^{-1}0.91 = 24.5°$.
$I_C = 12 \sin 60° - 6.593 \sin 24.5°$
$= 10.392 - 2.734 = 7.66$ A.

(b) $X_C = 240/7.66 = 31.33 \, \Omega = 1/100\pi C$,
and $C = 102 \, \mu F$.

1.11 (a) $X_L = 100\pi \times 100 \times 10^{-3} = 31.42 \, \Omega$.
$Z_L = \sqrt{(30^2 + 31.42^2)} \angle \tan^{-1}(31.42/30)$
$= 43.44 \angle 46.3° \, \Omega$.
$I_L = 200/ZL = 4.6 \angle -46.3°$ A.
$X_C = 144.7 \, \Omega$. $I_C = 200/144.7$
$= 1.38 \angle 90°$ A.
In-phase component of I_L
$= 4.6 \cos(-46.3°) = 3.178$ A.
Quadrature component of I_L
$= 4.6 \sin(-46.3°) = -3.326$ A.
$I = \sqrt{[3.1778^2 + (-3.326 + 1.38)^2]}$
$\angle \tan^{-1}(-1.945/3.178)$
$= 3.726 \angle -31.5°$ A.

(b) $Z = V/I = 53.68 \angle 31.5° \, \Omega$.
$Y = 1/Z = 18.63 \angle -31.5°$ mS.

(c) $P = I_L^2 \times 30 = 4.6^2 \times 30 = 634.8$ W.

2.1 (a) $V_{ph} = 240$ V. $I_L = I_{ph} = 240/48 = 5$ A.
(b) $P = \sqrt{3} \times 415 \times 5 = 3594$ W.

2.2 (a) $Z = \sqrt{(20^2 + 20^2)} \angle \tan^{-1}(20/20)$
$= 28.28 \angle 45° \, \Omega$.
$V_{ph} = 240$ V, $I_{ph} = I_L = 240/28.28 \angle 45°$
$= 8.49 \angle -45°$ A.
(b) $P = \sqrt{3} \times 415 \times 8.49 \times 0.707$
$= 4315$ W.

2.3 (a) $I_{ph} = 415/2200 = 0.189$ A.
$I_L = \sqrt{3} \times 0.189 = 0.327$ A.
(b) $P = \sqrt{3} \times 415 \times 0.327 = 235.1$ W.

2.4 $P = 15 \times 10^3 = \sqrt{3} \times 415 \times 0.7 I_L$.

$I_L = 29.81$ A.
$I_{ph} = 29.81/\sqrt{3} = 17.21$ A.

2.5 (a) $I_{ph} = I_L = 240/300 = 0.8$ A.
$P = \sqrt{3} \times 415 \times 0.8 = 575$ W.
(b) $I_{ph} = 415/300 = 1.383$ A,
$I_L = \sqrt{3} \times 1.383 = 2.395$ A.
$P = \sqrt{3} \times 415 \times 2.395 = 1721.5$ W.

2.6 (a) $P = \sqrt{3} \times 415 I_L \times 0.2 = 4500$.
$I_L = 31.3$ A.
(b) $V_{ph} = 240$ V. $|Z| = 240/31.3 = 7.67 \, \Omega$.
$\phi = \cos^{-1}0.2 = 78.5°$, $\tan\phi = X_L/R = 4.9$.
Therefore, $\sqrt{(R^2 + 4.9^2 R^2)} = 7.67$,
$58.83 = R^2 + 4.9^2 R^2 = 25.01 R^2$,
and so $R = \sqrt{(58.83/25.01)} = 1.53 \, \Omega$.
$X_L = 4.9 \times 1.53 = 7.5 \, \Omega = 100\pi L$,
and $L = 23.9$ mH.

2.7 $P_A + P_B = 10$ kW. $\phi = \cos^{-1}0.6 = 53.1°$,
so $\tan\phi = 1.33$.
$1.33 = \sqrt{3}[(P_A - P_B)/10\,000]$,
$P_A - P_B = (1.33 \times 10 \times 10^3)/\sqrt{3} = 7679$ W.
Therefore, $2P_A = 17\,679$, $P_A = 8839.5$ W.
$2P_B = 10\,000 - 7679 = 2321$, $P_B = 1160.5$ W.

2.8 The loads are $I_R = 240/100 = 2.4$ A,
$I_Y = 240/200 = 1.2$ A, and $I_B = 240/300$
= 0.8 A.
$I_N = 2.4 \angle 0° + 1.2 \angle -120° + 0.8 \angle -240°$
Total in-phase current
$= 2.4 + 1.2 \cos(-120°) + 0.8 \cos(-240°)$
$= 2.4 - 0.6 - 0.4 = 1.4$ A.
Total quadrature current
$= 1.2 \sin(-120°) + 0.8 \sin(-240°)$
$= -1.04 + 0.69 = -0.35$ A.
Therefore,
$I_N = \sqrt{(1.4^2 + 0.35^2)} \angle \tan^{-1}(-0.35/1.4)$
$= 1.44 \angle 14°$ A.

2.9 (a) $I_{ph} = 240/\sqrt{(200^2 + 1000^2)} \angle \tan^{-1}(1000/200)$
$= 0.235 \angle 78.7°$.
Volt-amps $= 240 \times 0.235 = 56.4$ VA.
Power $= 0.235^2 \times 200 = 11.05$ W.
(b) $I_{ph} = 240/400 = 0.6$ A.
Volt-amps $= 240 \times 0.6 = 144$ VA.
Power $= 0.36 \times 400 = 144$ W.
(c) $I_{ph} = 240/330 = 0.727$ A.
Power = volt-amps $= 240 \times 0.727$
$= 174.48$ W.
Total power $= 11.05 + 144 + 174.48$
$= 329.5$ W.
Total volt-amps $= 56.4 + 144 + 174.48$
$= 374.9$ VA.
Overall power factor $= 329.5/374.9 = 0.88$.

3.1 $n = \sqrt{(100/10)} = 3.162 : 1$.
$I = 1/20 = 0.05$ A.
$P_{L(max)} = 0.05^2 \times 10 = 25$ mW.

3.2 $V_{oc} = (10 \times 800)/1000 = 8$ V.
$R_{oc} = 240 + (800 \times 200)/1000 = 400 \, \Omega$.
$I_L = 8/(400 + 100) = 16$ mA.

3.3 $V_{oc} = (1 \times 1200)/1400 = 0.857$ V.
$R_{oc} = (1200 \times 200)1400 = 171.43 \, \Omega$.
$Z = \sqrt{(171.43^2 + 300^2)} \angle \tan^{-1} -300/171.43$
$= 345.53 \angle -60.3° \, \Omega$.
$I_L = 0.857/345.53 \angle -60.3°$
$= 2.48 \angle 60.3°$ mA.
$V_C = 2.48 \times 10^{-3} \angle 60.3° \times 300 \angle -90°$
$= 0.744 \angle -29.7°$ V.

3.4 $V_{oc} = (12 \times 22)/33 = 8$ V.
$R_{oc} = (22 \times 11)/33 = 7.33 \, \Omega$.
Hence $R_L = 7.33 \, \Omega$.
$I = 12/14.66$ A,
and $P_{L(max)} = (12/14.66)^2 \times 7.33 = 4.91$ W.

3.5 (a) $V_{oc} = 9 - [560(9 - 6)/860] = 7.046$ V.
$R_{oc} = (560 \times 300)/860 = 195.35 \, \Omega$.
$I_L = 7.046/(195.35 + 750) = 7.45$ mA.
(b) $I_{sc} = 6/300 + 9/560 = 36.07$ mA
and $R_{oc} = 195.35 \, \Omega$.
$I_L = (36.07 \times 195.35)/(195.35 + 750)$
$= 7.45$ mA.
(c) (i) Replace the 6 V source with a
short-circuit:
load on 9 V source
$= 560 + (300 \times 750)/1050 = 774.29 \, \Omega$.
$I'_L = (9/774.29) \times (300/1050) = 3.32$ mA.
(ii) Replace the 9 V source by a short-circuit:
load on the 6 V source
$= 300 + (560 \times 750)/1310 = 620.61 \, \Omega$.
$I''_L = (6/620.61) \times 560/1310 = 4.13$ mA.
Total load current $= 3.32 + 4.13$
$= 7.45$ mA.

3.6 (a) $n = \sqrt{(200/50)} = 2 : 1$.
(b) $I = 3/100$, $P_L = (3/100)^2 \times 50 = 45$ mW.
(c) $|Z_L| = \sqrt{(200^2 + 50^2)} = 206.16 \, \Omega$.
$n = \sqrt{(206.16/50)} = 2.03 : 1$.

3.7 The equation for the Thevenin equivalent circuit is
$V_L = V_{oc} - IR_{oc}$ and it is of the form
$y = mx + c$.
Thus R_{oc} is the slope of the curve and V_{oc} is the
constant. From the given figures, when $I_L = 12$ mA, $V_L = 2.4$ V, when $I_L = 6$ mA, $V_L = 7.2$ V.
Therefore, $R_{oc} = (7.2 - 2.4)/(12 - 6)$ kΩ
$= 800 \, \Omega$.
And $2.4 = V_{oc} - (12 \times 10^{-3} \times 800)$,
or $V_{oc} = 12$ V.
For the Norton circuit, $I = 12/800 = 15$ mA.

3.8 $X_C = 1/(2\pi \times 159 \times 1.25 \times 10^{-6}) = 800 \, \Omega$.
Hence, load impedance $= 600 \, \Omega$ resistance in
series with an inductor of 800 Ω reactance.
Therefore $L = 800/(2\pi \times 159) = 0.8$ H.
$P_{L(max)} = (4/1200)^2 \times 600 = 6.67$ mW.

3.9 (a) Load on source
$= [1000(1000 + 2200)]/4200 = 762 \, \Omega$.
$I_{in} = 6/762$ A. $I_{2.2} = (6/762) \times (1/4.2)$
$= 1.875$ mA.
$V_{oc} = 1.875 \times 10^{-3} \times 1000 = 1.875$ V.
$R_{oc} = (1000 \times 2200)/3200 = 687.5 \, \Omega$.
$P_L = (1.875/2)^2/687.5 = 1.278$ mW.

(b) With the output terminals short-circuited the
load on the 6 V source is
$(1000 \times 2200)/3200 = 687.5 \, \Omega$.
$I_{in} = 6/687.5$ A and
$I_{sc} = (6/687.5) \times (1/3.2) = 2.727$ mA.
$R_{oc} = 687.5 \, \Omega$. $I_L = 2.727/2$
and $P_L = (2.727/2)^2 \times 10^{-6} \times 687.5$
$= 1.278$ mW.

3.10 (a) $n = \sqrt{(390/2)} = 13.96 : 1$.
Reflected resistance $= 13.96^2 \times 2 = 390 \, \Omega$.
$I = 12/(2 \times 390)$, and
$P_L = [12/780]^2 \times 390 = 92.31$ mW.
(b) $348 = 42 + 2n^2$, $2n^2 = 306$,
$n = \sqrt{153} = 12.37 : 1$.
Reflected resistance $= 153 \times 2 = 306 \, \Omega$.
$I = 12/(348 + 42 + 306) = 17.24$ mA.
$P_L = (17.24 \times 10^{-3})^2 \times 306 = 90.96$ mW.

4.1 $R_p = V/I_1 = 20/(500 \times 10^{-6}) = 40$ kΩ.
$I_F = 400 \times 10^{-6} = V/R_1$,
$R_1 = 20/(400 \times 10^{-6}) = 50$ kΩ.
$40 = 50R_2/(50 + R_2)$, or R2 $= 200$ kΩ.
Time constant $= C(R_1 + R_2) = 2.5$ s.

4.2 $V_C = V_R = 40$ V.
$CR = 150 \times 10^3 \times 2.2 \times 10^{-6} = 330$ ms.
$v_R = 40 \, e^{-250/330} = 18.75$ V.

4.3 (a) Time constant $= L/R = 0.9/300 = 3$ ms.
(b) $I_1 = 0$.
(c) $I_F = 12/300 = 40$ mA.
(d) With the switch opened the time constant is
$0.9/1000 = 0.9$ ms and $I_1 = 40$ mA.
$i = 40 \, e^{-2.7/0.9} = 2$ mA.

4.4 (a) Time constant $= 0.1 \times 10^{-6} \times 500 \times 10^3$
$= 0.05$ s.
(b) $I_1 = 100/(1 \times 10^6) = 100 \, \mu A$.
(c) $I_F = 100/(2 \times 10^6) = 50 \, \mu A$.
(d) $V_C = I_F \times 1 \times 10^6 = 50$ V.
(e) $W = 0.1 \times 10^{-6} \times 50^2 \times 0.5 = 125 \, \mu$ J.

4.5 At 90% final amplitude, $0.9 \, V = V(1 - e^{-t/CR})$,
or $t_1 = 2.3CR$. At 10% final amplitude,
$0.1 \, V = V(1 - e^{-t/CR})$, or $t_2 = 0.1CR$.
Hence fall time $t_f = 2.2CR$ seconds.

4.6 Time constant $= 8/300 = 26.7$ ms.
$1/(\text{time constant}) = 37.5$.
$I_F = 60/300 = 200$ mA.
Hence, $31.6 = 200(1 - e^{-37.5t})$,
$0.158 = 1 - e^{-37.5t}$,
$\log_e 0.842 = -0.172 = -37.5t$,
and $t = 4.59$ ms.

4.7 $12 = 24 (1 - e^{-10/CR})$ (CR in milliseconds)
$0.5 = 1 - e^{-10/CR}$,
$\log_e 0.5 = -0.693 = -10/CR$,
$CR = 10/0.693 = 14.43$ ms,
and $C = (14.43 \times 10^{-3})/(100 \times 10^3)$
$= 0.14 \, \mu F$.

4.8 (a) $I_1 = 0$.
(b) $di/dt_{(t=0)} = V/L = 50/20 = 2.5$ A/s.
(c) $di/dt_{(i=0.2\,A)} = [50 - (0.2 \times 100)]/20$
$= 1.5$ A/s.

$(d) I_F = 50/100 = 0.5\,\text{A}$.

4.9 $W = 9.9 \times 10^{-3} = C \times 30^2/2$, and $C = 22\,\mu\text{F}$.
$5CR = 7.5\,\text{s}$. $CR = 1.5\,\text{s}$,
and $R = 1.5/(22 \times 10^{-6}) = 68.18\,\text{k}\Omega$.

4.10 $(a) > 100\,\text{ms}$.
$(b) < 1\,\text{ms}$.
(c) Cannot be done: the output should be from the capacitor.

5.3 $(a) P_{in} = \sqrt{(3)} V_L I_L \cos\phi =$
$\sqrt{3} \times 415 \times 4 \times 0.82 = 2.358\,\text{kW}$.
$(b) \eta = (2000/2358) \times 100\% = 84.83\%$.

5.4 Force on conductor $= BIl$ newtons. Total force F around armature $= NBIl$
$= 200 \times 50 \times 10^{-3} \times 4.5 \times 10 \times 10^{-2}$
$= 4.5\,\text{N}$.
Torque $= F \times$ armature radius
$= 4.5 \times 5 \times 10^{-2} = 0.225\,\text{N}\,\text{m}$.

5.5 Effective load on generator
$= 1/(1/150 + 1/25 + 1/120) = 18.18\,\Omega$.
Total circuit resistance $= 18.18 + 2 = 20.18\,\Omega$.
Terminal voltage $= (18.18/20.18)E = 0.9E$ volts.
Therefore, load power $P_L = (0.9E)^2/25$ and generated power $P_G = E^2/18.18$.
$\eta = (P_L/P_G) \times 100\% = 58.9\%$.

5.6 $N_1 = 600\,\text{r.p.m.}$ $I_L = 2000/300 = 6.67\,\text{A}$.
$I_f = 300/100 = 3\,\text{A}$.
$E_1 = 300 + (6.67 + 3) \times 0.25 = 302.42\,\text{V}$.
$I_L = 3000/300 = 10\,\text{A}$. $I_f = 3\,\text{A}$.
$E_2 = 300 + 13 \times 0.25 = 303.25\,\text{V}$.
Therefore, $E_1/N_1 = E_2/N_2$, and
$N_2 = (303.25 \times 600)/302.42 = 601.7\,\text{r.p.m.}$

5.7 $T \propto I_a^2$.
$(a) 50/I^2 = 25/25^2$, $I = 35.36\,\text{A}$.
$(b) 250 = E_1 + (25 \times 0.8)$, so that
$E_1 = 250 - 20 = 230\,\text{V}$.
$E_2 = 250 - (35.36 \times 0.8) = 221.71\,\text{V}$.
Therefore,
$230/(750 \times 25) = 221.71/35.36N$,
or $N = 511.2\,\text{r.p.m.}$

5.8 $P_o = 20 = 0.82P_i$, $P_i = 24.39\,\text{kW}$. Hence the power losses are $4390\,\text{W}$. Therefore friction/windage losses
$= 4390 - 3000 = 1390\,\text{W}$.

5.9 $300 = E + 0.1I_a = E + 10$, or $E = 290\,\text{V}$.
When the torque is doubled I_a is also doubled to $200\,\text{A}$ so that now $E = 280\,\text{V}$. Therefore,
$N = (1000 \times 280)/290 = 965.5\,\text{r.p.m.}$

6.1 $(a) 10 \log_{10}(12/1) = 10.79\,\text{dBm}$.
$(b) 10 \log_{10}(33/1) = 15.19\,\text{dBm}$.
$(c) 10 \log_{10}(1500/1) = 31.76\,\text{dBm}$.
(i) $13.79\,\text{dBm}$, $18.19\,\text{dBm}$, $34.76\,\text{dBm}$.
(ii) $7.79\,\text{dBm}$, $12.19\,\text{dBm}$, $28.76\,\text{dBm}$.

6.2 $P_{in} = (100 \times 10^{-3})^2/2000 = 5\,\mu\text{W}$.
$P_{out} = 2.5^2/1200 = 5.2\,\text{mW}$.
Gain $= 10 \log_{10}[(5.2 \times 10^{-3})/(5 \times 10^{-6})]$
$= 30.17\,\text{dB}$.
(Note that $20 \log_{10}(2.5/0.1) = 14\,\text{dB}$.)

6.3 $7-8\,\text{kHz}$.

6.4 Low-pass filter: $f_c = 7\,\text{kHz}$.
High-pass filter: $f_c = 4\,\text{kHz}$.

6.5 $9\,\text{dB} =$ voltage ratio of 2.82.
$R_1 = (600 \times 1.82)/3.82 = 286\,\Omega$.
$R_2 = (2 \times 600 \times 2.82)/(2.82^2 - 1)$
$= 487\,\Omega$.
(a) Overall gain $= 17 - 9 = 8\,\text{dB}$
$= 20 \log_{10}(V_{out}/V_{in})$,
$10^{0.4} = 2.51 = V_{out}/V_{in}$,
or $V_{out} = 2.51 \times 0.4 = 1\,\text{V}$.
$(b) P_{out} = 1/600 = 1.67\,\text{mW}$.
Therefore, $10 \log_{10}(1.67/1) = 2.23\,\text{dBm}$.
Alternatively, (b) $P_{in} = 0.4^2/600$
$= 2.67 \times 10^{-4}\,\text{W}$.
$8 = 10 \log_{10}[P_{out}/(2.67 \times 10^{-4})]$
$10^{0.8} = 6.31 = P_{out}/(2.67 \times 10^{-4})$, and
$P_{out} = 6.31 \times 2.67 \times 10^{-4} = 1.69\,\text{mW}$
$= 2.28\,\text{dBm}$.
$(c) V_{out} = \sqrt{(1.69 \times 10^{-3} \times 600)} = 1\,\text{V}$.

6.6 $(a) V_{out} = -30\,\text{dB w.r.t. } 1\,\text{mV}$
$= +30\,\text{dB w.r.t. } 1\,\mu\text{V}$.
$(b) 30 = 20 \log_{10}(1/V_{out})$
$10^{1.5} = 31.62 = 1/V_{out}$, or $V_{out} = 31.62\,\text{V}$.

6.7 (a) In decibels, $23.5\,\text{mW} = 10 \log_{10}23.5$
$= 13.7\,\text{dBm}$.
$1.25\,\text{W} = 10 \log_{10}[1.25/(1 \times 10^{-3})] =$
$31\,\text{dBm}$.
Fluctuation $= 31 - 13.7 = 17.3\,\text{dB}$.
Alternatively, fluctuation
$= 10 \log_{10}[1.25/(23.5 \times 10^{-3})]$
$= 17.3\,\text{dB}$.
(b) Maximum output power $= +11\,\text{dBm}$.
Minimum output power $= -6.3\,\text{dBm}$.
Fluctuation $= 11 - (-6.3) = 17.3\,\text{dB}$.

6.8 $|V_{out}/V_{in}| = 1/\sqrt{[1 + (f_c/f)^2]}$, where
$f_c = 3000\,\text{Hz}$.
For $6\,\text{dB}$ down, $|V_{out}/V_{in}| = \frac{1}{2}$
$= 1/\sqrt{[1 + (3000/f)^2]}$
$4 = 1 + 3000^2/f^2$,
or $f = \sqrt{3} \times 3000 = 5196\,\text{Hz}$.

6.9 $(a) 16 = 20 \log_{10}(V_{in}/1)$
$10^{0.8} = 6.31 = V_{in}/1$, $V_{in} = 6.31\,\text{mV}$.
$12 = 20 \log_{10}(6.31/V_{out})$,
$10^{0.6} = 3.98 = 6.31/V_{out}$, or $V_{out} = 1.59\,\text{mV}$.
$(b) P_{out} = (1.59 \times 10^{-3})^2/600 = 4.21\,\text{nW}$.

6.10 (a) (i) $100\,\text{W}$, (ii) $0.1\,\text{W}$, (iii) $2\,\text{mW}$.
(b) Gain of amplifier $= 20 \log_{10}26 = 28.3\,\text{dB}$.
Overall gain $= 28.3 - 10 = 18.3\,\text{dB}$
$= 20 \log_{10}(50/V_{in})$,
$10^{0.915} = 8.22 = 50/V_{in}$,
$V_{in} = 50/8.22 = 6.1\,\text{mV}$.

7.2 $(a) +127$, $(b) -32$, $(c) +7$, $(d) -10$.

7.3 $(a) 120\,\text{V}$.
$(b) 1\,\text{MHz}$.
$(c) 1500\,\text{Hz}$.
$(d) 0.6$.
$(e) 120 \times 0.6 = 72\,\text{V}$.

7.5   (a) $V_c + V_m = 75$, $V_c - V_m = 25$.
     Adding, $V_c = 50$ V.
  (b) $V_m = 75 - 50 = 25$ V.
  (c) $V_{SF} = 25/2 = 12.5$ V.
  (d) $m = 25/75 = 0.33$.

7.6   (a) $f_d = 500 \times 4 = 2000$ Hz
     $B = 2(2000 + 2500) = 9$ kHz.
  (b) $f_d = 1000$ Hz.
     $B = 2(1000 + 5000) = 12$ kHz.

7.7   (a) 10 kHz, 500 Hz.
  (b) $P_T = [(5/\sqrt{2})^2 \times 200] + [2 \times (1.5/\sqrt{2})^2 \times 200] = 2949$ W.
  (c) $\eta = (449.86/2949) \times 100 = 15.26\%$.

7.8 100 kHz $= 2(f_d + f_m) = 2(4f_m + f_m) = 10f_m$.
    Therefore, $f_m = 10$ kHz.
    6 dB is a voltage ratio of 2:
  (a) $m_f = 2 \times 4 = 8$ rad.
  (b) $B = 2(80 + 10) = 180$ kHz.

7.9   (a) $D = 15 = f_d/3400$, so $f_d = 51$ kHz.
  (b) $f_d = 51 \times 10/12 = 42.5$ kHz.
  (c) $f_d = 51 \times 6/12 = 25.5$ kHz.
     $m_f = 25.5/1 = 25.5$ rad.
  (d) $B = 2(51 + 3.4) = 108.8$ kHz.

7.10 (a) Signal is sampled four times per cycle at 50 μs, 175 μs, 300 μs and 425 μs. 4.7 V, 6.47 V, -4.7 V and -6.47 V.
  (b) —

8.1 $20 \times 50 \times 500 \times 4/1000 \times 1/2000 = 1$ m/V.

8.2 $V_g = 250$ V.
    $V_t = 249.1$ V $= K_A(V_{REF} - 249.1K_F) - 0.1$
    On 20 A load the internal voltage drop
    $= 20 \times 0.7 = 14$ V.
    $V_t = 250 - 14 = 236$ V.
    $V_t = kI_f - I_L r_a$ (A.1)
    $I_f = K_A(V_{REF} - K_F V_t)$ (A.2)
    When $I_L = 20$ A the increase in V_g needed
    $= 249.1 - 236 = 13.1$ V.
    From equation (A.1), $249.1 = kI_f - 14$,
    or $kI_f = 263.1$ V.
    Also $1 \times k = 250$ V.
    Therefore, $I_f = 263.1/250 = 1/0524$ A.
    Now, $1.0524 = K_A(85 - 249.1K_F)$ (A.3)
    On no-load, $V_t = V_g = 250$ V, $250 = kI_f$,
    $k = 250$ and $I_f = 1$ A. Hence
    $1 = K_A(85 - 250K_F)$ (A.4)
    Dividing equation (A.3) by equation (A.4) gives
    $1.0524 = (85 - 249.1K_F)/(85 - 250K_F)$
    Solving $K_F = 0.318$.
    Substituting for K_F into equation (A.4),
    $1 = K_A(85 - 79.5)$,
    or $K_A = 1/5.5 = 0.182$ A/V.

8.6 $0.7 = 50/F_c$, so $F_c = 50/0.7$
    $= 71.43$ N m per rad/s.

8.9   DE $=$ C4H $-$ 7AH $=$ 4AH.
    DFE $=$ 4AH/FFH $=$ 0.4AH.

9.1   (a) $C_x = 0.18/2 = 0.09$ μF $= 90$ nF.
     $r_x = 1500 \times 2 = 3000$ Ω.
  (b) $r_x = 3000$ Ω.
  (c) $X_L = X_C = 1/(2\pi \times 10^4 \times 0.18 \times 10^{-6})$
     $= 88.42$ Ω.
     $L_x = 88.42/(2\pi \times 10^4) = 1.4$ mH.

9.2   Clock frequency $= 10 \times (1/(20 \times 10^{-9})$
    $= 500$ MHz.

9.3   50 V r.m.s. $= 141.4$ V peak-to-peak.
    Vertical deflection $= 141.4/60 = 2.36$ cm.
    Horizontal deflection $= 260/65 = 4$ cm.

9.5   (a) Average $= 6.66/1.11 = 6$ V $=$ true r.m.s. value.
     % error $= [(6.66 - 6)/6] \times 100$
     $= 11\%$ high.
  (b) Peak $= \sqrt{2} \times 6.66 = 9.419$ V $=$ true r.m.s. value.
     % error $= [(6.66 - 9.419)/9.419] \times 100$
     $= 29.29\%$ low.

9.7 $C_x = 200 - 100 = 100$ pF.
    $1/82 = 1/85 + 1/Q_C$, $1/Q_C = 1/82 - 1/85$.
    Hence, $Q_C = 2323$.

9.9   Meter resistance $= 10 \times 3 = 30$ kΩ.
    The true voltage across AB
    $= (20 \times 100)/130 = 15.385$ V.
  (a) Total resistance BC $= 100$ kΩ || 30 kΩ
     $= 23.08$ kΩ.
     Indicated voltage $= (20 \times 23.08)/53.08$
     $= 8.696$ V.
     % error $= [(8.696 - 15.385)/15.385] \times 100 = -43.48\%$
  (b) Total resistance AB $= 30$ kΩ || 30 kΩ
     $= 15$ kΩ.
     Indicated voltage $= (20 \times 15)/115$
     $= 2.609$ V.
     Hence measured voltage BC $= 20 - 2.609$
     $= 17.391$ V.
     % error $= [(17.391 - 15.385)/15.385] \times 100 = +13.04\%$.

9.10 (a) 0.1% of 20.42 V ia 0.02 V. Hence, maximum voltage
     $= 20.44 + 0.02 = 20.46$ V
     minimum voltage
     $= 20.40 - 0.02 = 20.38$ V.
  (b) RCF $= -10 \log_{10}[1200/600] = -3.01$ dB.
     True reading $= +2.5 - 3.01 = -0.51$ dBm.

Index

a.c. bridge 191
  balance conditions 192
  capacitance 194
  commercial 195
  De Sauty 194
  frequency 195
  Hay 193
  Maxwell 193, 211
  Owen 194
  Schering 194, 211
  transformer ratio-arm 196
a.c. circuits
  admittance 7, 9, 10
  capacitance and inductance in
     parallel 12
  conductance 10
  impedance 1–4, 8, 9, 16, 18
  L and C in parallel 7
  parallel resonance 13, 19
  R and C in parallel 8
  R and C in series 2
  R and L in parallel 10
  R and L in series 1
  R, L and C in series 3
  reactance 1
  resonant frequency 8, 16
  series-parallel 14
  susceptance 7, 10
a.c. generator 87
  three-phase 28
accuracy, of a measurement 184
active filter 131
admittance 7, 9
alternator, motor car 89
amplitude modulation 134
  bandwidth 137
  instantaneous voltage 135, 155
  modulation factor 138
  modulation envelope 135
  overmodulation 141
  power carried 139
  sideband 138
  sidefrequency 137
apparent power 22
armature reaction 92
attenuation 119, 123
attenuation coefficient 124
attenuator 119, 125

bandwidth 18, 21, 137, 144, 147, 154

capacitance bridge 194
capacitive reactance 1
cathode ray oscilloscope (CRO)
  bandwidth of 204
  use to measure amplitude modulation
    203
    frequency 200
    phase 201
    risetime 203
CCITT 146
closed-loop control 158
conductance 10
commercial a.c. bridge 195
commutator 90
compound generator 97
computer control, of control system 173
control systems 156
  closed-loop 158
  computer/microprocessor control 173
  digital proportional control 176
  digital PI control 177
  digital PID control 178
  first order 160
  instability, of 179
  open-loop 156
  proportional control 160
  PI control 169
  PID control 170
  second order 160
  transfer function, of 166
current
  decay in RC circuit 62, 68
    time constant 63, 68, 70
  decay in RL circuit 74
    time constant 75
  growth in RL circuit 72
    time constant 73
current source 46

damping 180
damping ratio 181
data acquisition system 175
data output system 175
dBm 123
dBW 123
d.c. compound generator 97

d.c. generator 89
  armature reaction 92
  four-pole 92
  self-excited 94
d.c. motor 97
  series 99
  shunt 102
  universal 101
decibel 119
  reference levels 122
  voltage and current ratios 121
decibelmeter 189
delta connection 29, 36
demodulation 134, 145, 150, 151
derivative control 170, 178
De Sauty bridge 194
deviation ratio 143
differential phase shift modulation
    (DPSK) 147
differentiating circuit 80
digital control, of control systems 173
digital measurements 204
dynamic resistance 20

efficiency 86
e.m.f. constant 85
error, in measurement 184

fall time 76
filter 119
  active 131
  band-pass 129, 131
  band-stop 130
  high-pass 129, 131
  low-pass 126, 128, 131
first-order control system 159
first-order filter 126
four-pole d.c. generator 92
frequencies
  in AM wave 137–8
  in FM wave 143
frequency bridge 195
frequency deviation 142
frequency, effect on voltmeter 186
frequency modulation 142
  bandwidth 144
  deviation ratio 142
  frequency deviation 142

frequency swing 142
  modulation index 142
  narrow band (NBFM) 144
  rated system deviation 143
signal-to-noise ratio 145
frequency, resonant 8, 16, 19
frequency shift modulation (FSK) 146
frequency swing 142

generator
  a.c. 87
  d.c. 89
    compound 97
    series 96
    shunt 97
generator effect 85

Hay bridge 193

impedance 1–4, 8, 9, 16, 18, 20
induction motor 105
  rotating magnetic field 106, 110
  shaded pole 115
  slip 112
  speed of field 113
  split phase 115
inductive reactance 1
instability, in a control system 179
integrating circuit 78
integral control 169, 177

line current 32, 36
line voltage 33, 36
Lissajous' figures 200
loading effect, of voltmeter 185
logic analyser 206
  state analysis 209
  timing analysis 207
logic checker 205
logic probe 205
logic pulser 205

maximum power transfer theorem 57
Maxwell bridge 193
measurement,
  of amplitude modulation 203
    decibels 189
    frequency 200
    phase 201
    risetime 203
    three-phase power 41
    voltage 184
microprocessor control, of control
        system 173
modulation 132
  amplitude 134
  frequency 142
  phase 147
  pulse 148
modulation envelope 135
modulation factor 138
modulation index 142
motor car alternator 89

motor, d.c. 97
  braking 103
  shunt 102
  series 99
    starting 103
    universal 101
  induction 105, 110, 114
  small 114
  stepper 115
motor speed control 100, 102, 163

narrow-band frequency modulation
        (NBFM) 144
National grid 38
network theorems
  maximum power transfer 57
  Norton 50
  superposition 54
  Thevenin 47
neutral current 39
neutral point 29
Norton's theorem 50

open-loop control 156
overmodulation 141
Owen bridge 194

parallel resonance 13, 19
  bandwidth 21
  dynamic resistance 20
  Q factor 20
  resonant frequency 19
  selectivity 21
phase current 33, 36
phase modulation 147
phase sequence 27
phase shift modulation (PSK) 147
phase voltage 27, 30, 36
power
  apparent 22
  three-phase 34, 39
  true 22
power factor 22, 39
power factor correction 23
principle of generator/motor 84
proportional control 160, 176
proportional plus integral control
        169, 177
proportional plus integral plus derivative
        control 170, 178
pulse amplitude modulation (PAM) 148
pulse code modulation (PCM) 151
  bandwidth 154
  pulse regeneration 154
  quantization 152
  time–division multiplex 153
pulse duration modulation (PDM) 150
pulse position modulation (PPM) 151
pulse repetition frequency 76
pulse waveform 75
pulse width or duration 76

quadrature amplitude modulation
    (QAM) 148

Q factor 16
Q meter 199

rated system deviation 143
ratio arm bridge 196
reactance 1
remote position control 161
resolution, of measurement 183
resonance
  parallel 13, 19
  series 7, 15
resonant frequency 8, 16, 19
risetime 76, 203
rotating magnetic field 106

Schering bridge 194
second-order control system 159
second-order filter 128
selectivity 17, 21
self-excited d.c. generator 94
sensitivity, of a measurement 183
series d.c. generator 96
series d.c. motor 99
series d.c. shunt motor 102
series-parallel circuit 14
series resonance 7, 15
  bandwidth 18
  Q factor 16
  resonant frequency 16
  selectivity 17
  voltage magnification 16
shaded-pole induction motor 115
shunt d.c. generator 95
shunt d.c. motor 102
sideband 138
sidefrequency 137, 144
signature analyser 210
slip 113
small electric motors 114
speed control, of a motor 100, 102, 163
split-phase induction motor 115
star connection 29, 31
star point 29
stepper motor 115
superposition theorem 54
susceptance 7, 10

Thevenin's theorem 47
three-phase circuits 27
  delta connection 28, 36
  line current 33, 36
  line voltage 30, 32, 36
  measurement of power 41
  measurement of power factor 43
  neutral current 39
  phase current 33, 36
  phase sequence 27
  phase voltage 27, 30, 36
  power dissipated 34, 39
  star connection 29, 31
  star point 29
time constant 63, 73
time-division multiplex 153
torque constant 85

transfer function 166
transformer ratio arm bridge 196
true power 22

universal motor 101

voltage, growth in a RC circuit 62

voltage magnification 16
voltage regulation system 164
voltage source 46
voltmeter
  frequency effect 186
  loading effect 184
  waveform effect 187

water level control 160
waveform error, in voltmeter
  average responding 188
  peak responding 189
Wheatstone bridge 184